An Innovative Solution for Environmental Issues

By: PWPA Los Angeles

Maiko Shimogawara

Jasmine Santoro, Jinka Kawasaki,

Akane Horii, Yuta Horii

Table of Contents

Preface

In the 20th and 21st centuries, scholars and scientists have taken a significant role in helping humanity create remarkable inventions in numerous fields, enabling us to enjoy a life of abundance. However, many of these inventions have come with unforeseen consequences, some of which pose serious threats to human life and the well-being of all living things. Without resolving these environmental issues, the Earth's environment and human civilization may not have a future. In this solution guide, we present a holistic approach to address environmental issues, one that can unlock nature's potential to heal the Earth. Our inspiration for this solution guide comes from Rev. Dr. Sun Myung Moon and Dr. Hak Ja Han Moon, who believed that scientific achievements need to address problems in human history in order to fulfill the ideals of peace, happiness, and harmony with the environment.

Rev. Dr. and Dr. Moon founded and established Professors World Peace Academy (PWPA) International in 1973 with the vision to mobilize intellectuals who have devoted their lives to the advancement of human wisdom and enable them to play an important role in overcoming the dangers within society and opening new pathways to world peace. Another initiative started by Rev. Dr. Sun Myung Moon and Dr. Hak Ja Han Moon is the International Conference on the Unity of Sciences (ICUS), an interdisciplinary academic forum that examines global issues. Established in 1972, it was recently revived in 2017 with the participation of scholars from across the globe. PWPA has developed in 83 nations, holding over 300 international conferences, with the attendance of Nobel Prize laureates such as Lord Edgar Adrian, Dr. Sir John Eccles, Dr. Eugene Wigner, and Dr. Friedrich von Hayek. Notably, Dr. Morton Kaplan, a Distinguished Service Professor at the University of Chicago, served as the President for over 30 years since its founding. ICUS has organized 28 international conferences, drawing the participation of Nobel Prize laureates Dr. Shuji Nakamura, Dr. David MacMillan, and Dr. Mario Molina. The goal of ICUS, like PWPA, is to help create a just and harmonious world order by encouraging scientists to search for new ideas and methods of achieving peace and prosperity. Since the establishment of PWPA and ICUS in the 1970s, Rev. and Dr. Moon have brought together conscientious scientists and scholars to address the world's complex challenges. They recognized that solving these problems requires a holistic, universal approach. Central to their vision was the belief that absolute values must be at the center of order in order to unify science, morality, and philosophy, for the achievement of authentic human happiness and lasting peace.

Inspired by the vision of the PWPA founders, several local youths revived the PWPA Los Angeles Chapter in June 2022. Since then, the PWPA-LA has organized six in-person conferences and ten online forums, launched a website, and continuously delivers projects in collaboration with scholars. We received tremendous support from Dr. Gordon Anderson, Secretary General of PWPA, Dr. Frank Kaufmann, President of PWPA, and especially from Dr. Thomas and Grace Selover, the current President of PWPA International. We hope to be a platform where conscientious professors can share their expertise anchored in absolute values, which are currently difficult to convey within the existing educational systems and make this knowledge accessible to all people across the globe.

The development of civilization has wrought detriment to the environment. Countless wars and conflicts among nations continue to destroy human lives and accelerate damage to our planet Earth. When we look at the realities afflicting the world today, we feel that the future is bleak without addressing these challenges. Dr. Hak Ja Han Moon highlights the urgency of our times, stating, "If the planet has no future, humankind has no future. This is the direction we are heading." At the same time, she finds great hope in all scientists and scholars' efforts and their dedication to correct this direction. Mutual cooperation beyond national and regional boundaries and interdisciplinary study beyond limited specializations are absolutely necessary. The dedication of each scholar's research and hard work is the hope for the future and humankind. It will become the joy of all created things and, through that effort, bring joy, delight, and glory to God, our Creator. In doing so, a new age and a new history will unfold.

She is asking for your active cooperation and effort. Together with Dr. Moon, members of PWPA-LA believe that we can find a way to correct what was distorted through the development of unprincipled science.

In this solution guide, we present a holistic approach to address environmental issues. We sincerely hope this effort can spark a ripple effect, inspiring meaningful changes on a bigger scale. Dr. Moon once gave the example of a monarch butterfly. They are small, only about three centimeters long. With such small movements, they fly annually all the way from Canada to Mexico. Who told them to go in that direction? These butterflies will go through many difficulties and will have to make many sacrifices. Nevertheless, they will go. In the same spirit, we hope to complete the journey forward, overcoming those difficulties on the way together until we can see the bright future that we can leave for future generations.

We would like to express our sincere gratitude to all those who contributed to the publication of this solution guide. We are deeply grateful to the PWPA-LA advisory committee members, Dr. JoAnna Schilling, President Emeritus of Cypress Community College, Dr. David Blekhman of California State University Los Angeles, and Professor Grace Apiafi of Pasadena City College for their unwavering support and guidance. We extend our heartfelt gratitude to Dr. Rattan Lal of Ohio State University, Dr. William Keepin of Satyana Intsitute, Dr. Cliff Davidson of

Syracuse University, Dr. Michael Stenstrom of University of California, Los Angeles, and Dr. Andrew Combs of the World Research Institute for Science and Technology for contributing their extensive research and knowledge. Additionally, this publication would not have been possible without the insightful advice, help, and wisdom of Dr. Gordon Anderson of PWPA and Dr. Takahiro Hiroi of Brown University..

We offer this book, *An Innovative Solution for Environmental Issues*, to all, along with the heartfelt intentions of our founders, Dr. Sun Myung Moon and Dr. Hak Ja Han Moon, who sought to share their love and hope with the world. While this work may not yet be complete, it is our sincere hope that it will inspire each reader in a unique and meaningful way. We look forward to continuing this journey together with everyone reading this book today.

Professors World Peace Academy Los Angeles

March 31, 2025

Chapter 1

New Approach

The creation of the wheel was one of the most revolutionary inventions in human history. With its development, agriculture flourished, transportation became more efficient, and it inspired transformative changes in manufacturing processes. Today, the wheel can be found everywhere, from the tires on a vehicle to the rotating parts of the engine. In all levels of production and fields, the wheel plays an invaluable role.

Humanity has progressed significantly from the simplest of inventions into the modern world. The history of science and technology, closely intertwined with human history, has often propelled improvements in the quality of life. This raises an important question: What exactly has been the nature of the history of science and technology, and how does understanding this history contribute to finding the solutions for the environmental crisis faced today?

History of Development in Science and Technology

The history of science and technology has developed in parallel with human history. There are many moments where technological advancements surpassed what was once deemed possible, ushering in a time when the quality of life could be significantly improved. The history of science and technology has largely been shaped by a desire to understand nature to improve the quality of life – a journey that focused on harnessing the power of nature to meet human needs. Advancements in the understanding of science and technology played an integral role in improving human lives. However, the foundation of this progress has been rooted in a tendency to manipulate or use nature as a tool to fulfill humans' immediate needs rather than to live sustainably with the environment.

With the development of technological advancements, little consideration was given to how these efforts may affect the balance and health of the environment. This method of development has invariably led to the environmental challenges seen today, such as air pollution, resource depletion, damaged soil, and many other issues. These issues are not the result of malice but of a limited understanding of the importance of working in harmony with nature.

As a result, the relationship between humans and nature has come to resemble a form of conflict and domination. Historical trends indicate that human beings have not been effective stewards of nature. There have not been many moments where humanity reflected on its treatment of the environment or sought to improve its care for the natural world. In fact, by turning towards the thesis-antithesis-synthesis model (Figure 1.1) of conflict within nature as an example, the conflict of science was used as a means of justifying the type of relationship humans share with nature. As a result, humans have become accustomed to adopting the idea that everything in the world develops through a series of conflicts, even when they relate to the environment as well. However, this closes the door to development through harmonious and cooperative relationships with nature.

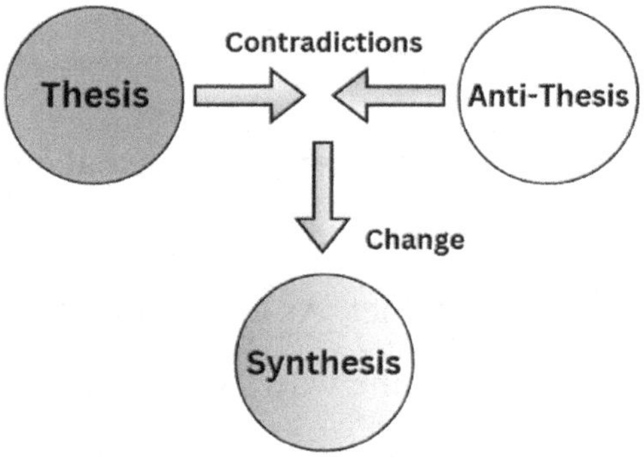

Figure 1.1 Thesis-Antithesis-Synthesis Model of conflict. Created by the author.

It is true that in recent years, there has been a large push to protect nature. The rise of non-profit organizations, international conferences, and the growth of renewable energy industries are evidence of this shift. However, there is still significant room for improvement. The environmental crisis remains an urgent and pressing concern, requiring an even greater effort to address the challenges it presents.

Thus, a new approach is needed – one centered on understanding and learning from nature. Rather than interfering with and attempting to control it, humanity must adopt a mindset of working in harmony with the natural world. This shift will unlock the potential of nature, and in turn, help create a better world. Humanity and the natural world should share a symbiotic relationship where both help each other.

The World Continues to Face Pressing Issues

Scholars have consistently warned about the urgent need for immediate action to address the environmental challenges facing the planet. As a response, humankind has made efforts to improve the way we interact with nature, even on a global scale. Concern for the natural environment is a sedimented idea. However, it remains true that despite widespread efforts to protect nature, progress falls short. Why is this the case? Is the problem rooted in the way nations interact with one another? Is it a matter of technological inadequacy? Or is the lack of innovative solutions the primary obstacle? The answer to these questions cannot be reduced to a single explanation. A comprehensive response is needed, which examines the issue from multiple angles. However, one thing is clear: technology is not the limiting factor in resolving the world's environmental crisis.

As a matter of fact, technology has advanced, and yet the same environmental issues persist. New inventions and breakthrough research are being driven by scholars who have invested many years with hopes of creating a better world, yet it still struggles to make a global impact.

Look at the current allocation of resources spent on agriculture versus global food security. 38% of the Earth's terrestrial surface is used for agriculture, 75% of agricultural land is allocated to raising animals, and 70% of the global freshwater withdrawals are used for irrigation (Lal, 2022). Despite this, 1 in 7 persons are food insecure, and the global hunger index continues to grow with every passing year. The current food system is not working. However, this failure is not due to a lack of funds or available land for agriculture. Research indicates that the resources currently available should be sufficient. Rather, the problem lies in the model of how humanity interacts with soil and agriculture. The solution must address this fundamental problem. Dr. Rattan Lal, a Professor of Soil Science and Director of the Rattan Lal Center for Carbon Management and Sequestration at Ohio State University, perfectly captures this idea by emphasizing the need to transition from taking land from nature to returning land back to nature.

Consider the issue of air pollution. Air pollution is a significant environmental concern, costing billions of dollars globally in health and welfare impacts. Yet, the technological solutions to this issue largely involve redesigning existing energy systems to rely on renewable sources such as solar power – technology that is already available. For instance, redesigning an electric water heater into a solar water heater is an idea and resource that is already available. It simply needs to be redesigned to meet the specific need. Once again, this reinforces the point that the necessary technology and innovation already exist. However, it is the other limiting factors that are preventing a full transition or an effective utilization of these resources.

Offering a New Approach

It all boils down to this. Technology has advanced to a point where it can address many of the environmental challenges the world faces. The required tools and resources exist, and they are not the limiting factor in the effort to protect nature. The fundamental problem lies in the absence of the right mind structures and philosophies guiding the development and use of technology. As a result, even with the right values and social institutions to guide the development, meaningful action remains difficult. It is true when they say a tool is only as effective as the person who wields it. Humanity has long adhered to a philosophy of dominating and controlling nature, a mindset that inadvertently influenced the way solutions are pursued. For much of human history, humanity has interfered with the natural world processes rather than understanding and respecting it. However, if humanity is able to grasp the principles and order of nature, it will be possible to harmonize with the natural world and realize effective solutions to the environmental crisis.

What exactly is the new approach that will enable humanity to harmonize with nature and unlock its full potential? What kind of shift in understanding must happen? Historically, the relationship between technology, science, and humanity has been shaped by a philosophy of dominating and controlling nature. This mindset has hindered the development of adequate solutions even in modern times. The new approach focuses on transitioning from a self-centered dominion over nature to a stewardship rooted in love. Achieving this requires understanding the hidden order within the natural world. Humanity must learn to live in alignment with a higher purpose.

Everything is connected. From the particles that make up matter to the vast universe, all are part of a whole that aligns with a central purpose. Unification Thought[1] proposes the idea that everything is interconnected within a system in the created world. At the simplest level, elementary particles maintain their existence as particles, but they exist to form atoms. Similarly, cells sustain their existence and function as cells, but they serve the purpose of forming tissues and organs. The Earth, while maintaining its existence, plays a role in forming the solar system. From the smallest particle to the largest planet, each level works together to create one unified, organic body.

The phenomenon of quantum entanglement best demonstrates the potential of interconnectedness, where

---

[1] Book published in English in 1973, based on a system of thought developed by Rev. Sun Myung Moon which seeks to unite diverse ideologies by overcoming materialism, humanism, and religious divisions. It emphasizes universal principles, underlying laws of nature, and the moral foundation of society, offering a framework for achieving peace and unity.

two or more particles are linked in a way that transcends time and space. This suggests that particles do not exist in isolation, pointing to the possibility of a larger interconnected system. Beyond energy and matter, there is a sense of consciousness that binds all things together, offering a holistic perspective.

With all of this in mind, the question arises: Where do humans stand within this higher order of all things? How can human beings align with the natural order? By understanding this alignment, humanity can unlock nature's potential and accelerate efforts to address the environmental crisis using current technologies.

Chapter 2

Addressing Human-Caused Environmental Issues

Before seeking solutions, this chapter will explore the current realities we face. Human behaviors, driven by the desire for a more comfortable lifestyle, have led to numerous consequences. Many of these are now manifesting as environmental issues that impact humanity itself. Among many challenges, this chapter will focus on three key issues: energy resources, water pollution, and soil and agronomy. We will examine the human actions behind these consequences, review the technologies and research being used to address these problems, and offer projections for the future if we continue our way of life based on expert research.

Section 1

Environmental Impacts of Energy Production

Written by: Jasmine Santoro

Energy has become an essential worldwide, powering many aspects of life and technology. The greater need for energy brings many economic and environmental concerns and challenges. The discussion on energy can be complex, so this section will provide only a brief overview of the challenges in energy production and the urgent need for innovative solutions.

Worldwide, an estimated 760 million people live without access to electricity needs at home, and 2.3 billion live without access to clean cooking (International Energy Agency [IEA], 2023). Despite this imbalance of energy distribution, oil production and consumption hit record highs in 2023 (Energy Institute, 2024). Until now, energy production has relied heavily on fossil fuels, but the environmental consequences are becoming impossible to ignore. Carbon dioxide emissions have tripled since the 1960s, primarily because of fossil fuels (National Oceanic and Atmospheric Administration [NOAA], 2024). There are also limitations to the finite supply of fossil fuels as they are becoming increasingly difficult to extract (Blekhman, 2023).

Significant work has been done to develop renewable sources of energy. However, the transition from fossil fuels has been slow due to political and economic factors involved in the global energy trade. As the environmental harm of fossil fuels, and ultimately, human-caused climate change worsen, the need for innovation in energy production is more urgent than ever.

Our Reliance on Fossil Fuels

Our dependence on fossil fuels has continued to grow since the Industrial Revolution. Beginning with coal, it later expanded to include natural gas and oil. Because of this, large-scale infrastructure for energy production has been centered on fossil fuels. In 2023, these three sources generated 60% of electricity globally (Energy Institute, 2024).

In fossil fuel combustion, they are burned, which generates usable energy. Their high carbon content reacts with the oxygen and produces greenhouse gases, primarily carbon dioxide, methane, and nitrous oxide. While these gases exist naturally to maintain Earth's warmer, livable temperature, excessive emissions from human activities have led to human-caused climate change (National Aeronautics and Space Administration, 2024). In 2022, fossil fuel combustion for transportation led to 35% of CO_2 emissions in the United States. Electricity accounted for 30% of the total, and industrial processes accounted for 16% (Environmental Protection Agency [EPA], 2025a).

Despite documented environmental damage, the transition away from fossil fuels has been slow. The fossil fuel industry remains profitable at the cost of sustainability, and corporations and governments have benefited economically. In 2022, global investments in technology for cleaner energy passed $1 trillion (USD), yet investments in fossil fuels were nearly double that (International Renewable Energy Agency, 2023). That same year, profits from oil and gas companies reached a record $916 billion (USD), with the United States gaining over $300 billion (USD). Much of it went to shareholders (Semieniuk et al., 2024).

Transitioning from fossil fuels comes with economic and infrastructural challenges. The current energy system has been built around fossil fuels and remains profitable to an extent. Moving to large-scale renewable energy will require a high upfront investment, but in the long run, it is expected to be more cost-effective (IEA, 2021).

Why We Should be Worried about Fossil Fuels

On its own, the Earth has been able to naturally balance its carbon through its carbon cycle (Figure 2.1). This is the cycle of carbon between the atmosphere, Earth's interior and surface, including plant and animal life, and the ocean. Some carbon is stored in the Earth's interior in rocks and sediments and is released through volcanic and geologic activity. As it enters the atmosphere, it exists primarily as carbon dioxide and methane. These are gases that are absorbed by plants for photosynthesis. On Earth's surface, there is an exchange of carbon between plants, animals, and the soil, where decomposers further release carbon back into the environment. The ocean has a vital role in the carbon cycle as it absorbs large amounts of atmospheric carbon (US Department of Energy, n.d).

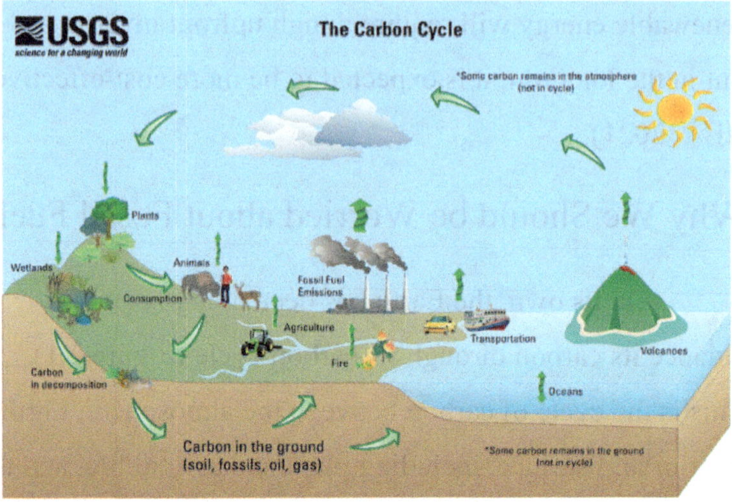

***Figure 2.1** the Carbon Cycle*

Note. Reprinted from US Geological Survey. Public domain.

While this cycle has long helped Earth maintain equilibrium, human activities have disrupted this balance. The extraction and burning of fossil fuels have released significant amounts of carbon into the atmosphere faster than the cycle can balance. These human-caused emissions have tripled from 10.9 billion tons to 36.8 billion tons over the last 65 years (NOAA, 2024). Fossil fuels accounted for almost 90% of carbon dioxide emissions (United Nations, n.d.). Coal-fired power plants produce 2.2 to 2.7 pounds of carbon dioxide for every kilowatt-hour (kWh) energy generated. For natural gas, it is 0.9 to 1.2 pounds, and for oil-fired power plants, it is 2.0 to 2.7 pounds (Figure 2.2) (Al Mubarak et al., 2024).

Fossil-fuel based power plant	**Coal**	**Natural Gas**	**Oil**
Pounds of CO_2 for every kWh of energy	2.2 - 2.7	0.9 - 1.2	2.0 - 2.7

Figure 2.2 *Pounds of carbon dioxide produced for every kilowatt-hour energy generated for each fossil-fuel based plant*

Note. Data from Al Mubarak et al. 2024. *Economic, societal, and environmental impacts of available energy sources: A Review.*

Human activities have destabilized the carbon cycle, causing an unprecedented accumulation of anthropogenic carbon in the atmosphere. Carbon dioxide is particularly alarming because if not absorbed in other parts of the carbon cycle, it can persist in the atmosphere for thousands of years. Nitrous oxide's atmospheric lifetime is over hundreds of years, and methane's is a little over a decade (EPA, 2025b). While fluctuations in atmospheric carbon have occurred naturally in Earth's history, the steady rise since the Industrial Revolution has led to significantly higher levels today. Compared to pre-industrial times, atmospheric carbon has increased by over 50%, reaching 419 ppm (parts per million) in 2023. Methane levels have risen by more than 160%, reaching 1,922 ppb (parts per billion), and nitrous oxide has increased by over 25%, reaching 336 ppb (Figure 2.3) (NOAA, 2024).

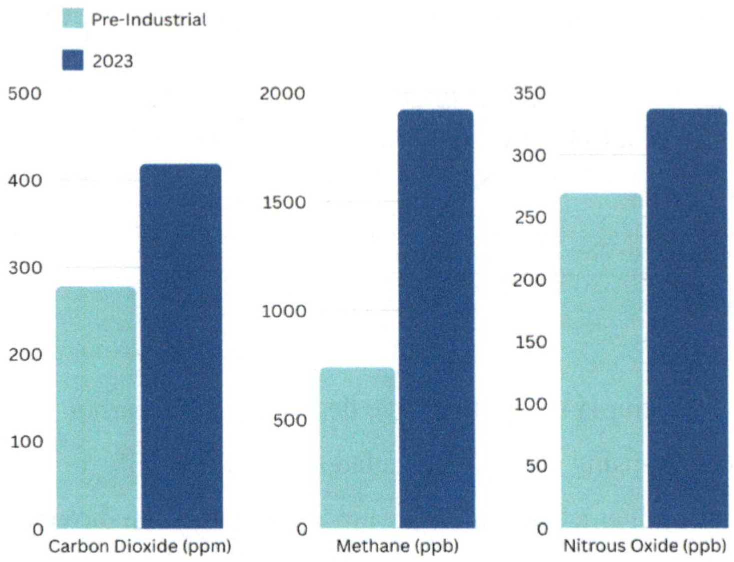

***Figure 2.3** Greenhouse gas levels in pre-industrial times in comparison to 2023*

Note. Data from National Oceanic and Atmospheric Administration. 2024. *No sign of greenhouse gases increases slowing in 2023.*

This directly contributes to human-caused climate change, resulting in more severe and frequent weather events, like droughts, floods, and wildfires. It has also led to an increase in overall global temperature. The average temperature in the 20th century was 13.9°C. In 2023, the average temperature increased by 1.18°C, making it the hottest year on record. This is not an anomaly, as the past ten years have been the ten warmest since global record-keeping began in 1850 (NOAA National Centers for Environmental Information, 2024). In the United States alone, over $605 billion (USD) was spent on handling extreme weather events in the four years between 2016 and 2020 (Bertrand, 2021).

Beyond climate effects, the large amounts of greenhouse gases contribute to acid rain and ocean acidification. This harms both terrestrial and marine ecosystems. The ocean absorbs over 25% of atmospheric carbon, increasing its average acidity by 30% over the last 150 years. This higher acidity reduces carbonate, a substance shellfish needs to form their shells and skeletons. If the acidity continues to rise, the ocean could reach a point where it dissolves these shellfish's shells. This would affect thousands of species and the marine ecosystem (Natural Resources Defense Council, 2009).

The emissions also have direct consequences on human health. Air pollution from fossil fuels is a major contributor to respiratory diseases, heart conditions, and cancer. It is estimated in 2018 that one in five (8 million) deaths were related to fossil fuel pollution (Miller, 2021). So, while projections warn of worsening impacts, the damage is already evident today.

Niger Delta as an Example

An area that has been devastated by fossil fuels is the Niger Delta Region, Nigeria's center for oil wealth. Since the discovery of oil in the 1950s, the region has experienced political unrest and significant environmental degradation (United Nations Environment Programme, n.d.). The oil companies here have neglected pipeline maintenance. This, combined with vandalism and attempts at oil theft, has led to frequent oil spills that have severely damaged the environment and local communities (Allen, 2012). Once a region of rich biodiversity, it has become plagued by pollution. Nearly 7,000 oil spills have been recorded between 1976 and 2001, amounting to 1.8 million barrels of oil contaminating the land and water. These spills continue to this day, creating a disaster that some argue surpasses the devastation of BP's Deepwater Horizon spill in 2010 (Aka, 2024). The region also has high amounts of gas flaring. The local communities bear the harmful effects as their environment, including farmlands, marine life, and animal and plant species important to their livelihood, are severely endangered (Ebiede, 2011). Studies have also linked the oil spills to rising neonatal deaths in the region (Bruederle & Hodler, 2019).

Since the nation and government have benefitted from the oil revenue, it became difficult for them to prioritize resolving the damages done to the environment and local populations. Dr. Ibaba writes that the accumulation of wealth from the Niger Delta to the government leads them to neglect the public good, leading to corruption, lack of development in the region, and human rights violations (Ibaba, 2011).

Renewable Energy

Efforts have been made to transition from fossil fuels to renewable energy sources such as solar, wind, hydrogen, and nuclear power. While these alternatives offer significant potential, the widely available technology today still faces some challenges, including reliability, large land requirements, and high initial costs. Meeting global energy demands requires high upfront investments, which has slowed large-scale infrastructure for renewables (IEA, 2021).

Solar and wind are projected to grow faster than other energy sources, but their intermittent nature poses a challenge. Sunlight and wind are not constant, making their energy generation less predictable, and this is also highly dependent on the region of the world in which it is generated (Blekhman, 2023; US Energy Information Administration [EIA], 2023). Scaling up these technologies requires vast amounts of land, raising concerns about their impact on the surrounding ecosystems, as they often force out and harm wildlife in the area (Badichek, 2019).

Solar panel usage is rapidly increasing, but concerns about solar panel materials have arisen. These include the corrosive chemicals in solar panel production and their waste management. Estimates suggest that in 15 to 30 years, up to 100 million solar panels could become waste. As they contain toxic materials such as cadmium, lead, and arsenic, their disposal has an environmental impact (Hamilton, 2015). However, research on how to overcome these issues continues today and may be resolved in the future.

Looking Forward

As mentioned earlier, oil production and consumption reached record levels in 2023, with over 96 million barrels produced daily and over 100 million barrels consumed daily. This has led to an all-time high of energy-related greenhouse gas emissions (> 40GtCO2e) (Energy Institute, 2024).

Projections say that the use of coal, the largest carbon dioxide emitter, is expected to decrease, along with emission intensity. However, in almost all projections, carbon dioxide emissions are expected to increase from now. Natural gas is projected to increase anywhere from 11 to 57%. Global transportation demand is projected to rise, further increasing energy consumption by 8% to 41% (depending on the projection model used) (US EIA, 2023). In a world already experiencing the detrimental effects of fossil fuels, this trend raises a critical question: What will happen if we do not make any changes?

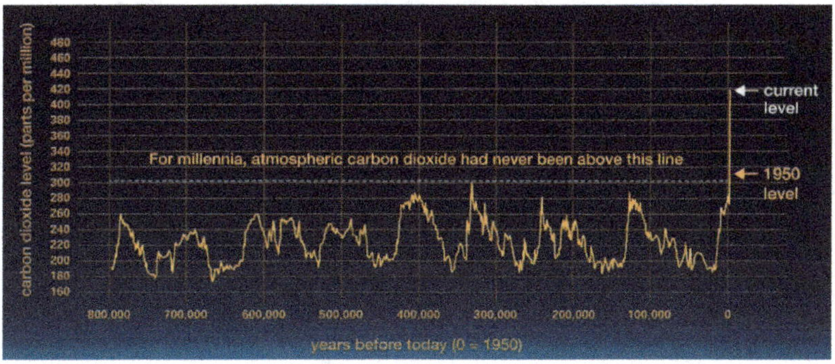

Figure 2.4 *CO_2 levels (ppm) from 875,000 years ago until data in 2008. In 2023, we are at an unprecedented level of 419 ppm.*

Note. Data from Luthi, D., et al.. 2008; Etheridge, D.M., et al. 2010; Vostok ice core data/J.R. Petit et al.; NOAA Mauna Loa CO2 record. Some description adapted from the Scripps CO2 Program website, "Keeling Curve Lessons." Graphic from NASA (National Aeronautics and Space Administration), 2023. Public domain.

At this rate, atmospheric carbon dioxide will continue to rise as it has been (Figure 2.4), making the future goal of net zero insufficient to combat rising temperatures (>1.5°C) (Intergovernmental Panel on Climate Change, 2023). Many regions already suffer from climate-driven disasters, including flooding, prolonged droughts, and escalating wildfires. Rising sea levels, which have increased by 8-9 inches since 1880, are projected to worsen (Lindsey, 2023). Without changes, these crises will only intensify.

There is hope for sustainable energy production through hydrogen and nuclear (small modular reactors) energy. These emerging technologies have the potential to provide stable, low-carbon energy, though they require significant infrastructure investment and continued safety improvements.

The innovation of energy resources has transformed the way people live and has become essential for other technologies. We cannot deny that energy-generating technology has improved our quality of life and enabled greater productivity despite the environmental damage it has caused. These benefits should be accessible to everyone. Considering the many populations still lacking access to electricity and energy resources, global energy consumption will continue to rise. At the same time, to ensure a pleasant living environment for all humans and lives on Earth, we must restore the natural world we have destroyed. We urgently need innovative solutions to meet the rising global energy demand while protecting our planet. By advancing energy production and consumption innovations, we can redefine our relationship with nature and work toward a more sustainable future.

Section 2

How City Infrastructure Contributes to Water Pollution

Written by: Akane Horii

Among the variety of global environmental issues humanity faces, water pollution remains a pressing issue. In correlation with the development of industrialization and human population growth, water quality near large water bodies is deteriorating at an alarming rate (Zutshi & Prasad, 2008). Wastewater discharges have deposited sludge blankets over a meter thick on the ocean floor in many regions (Stenstrom, 1998), and the survival of many marine species is threatened as their populations decline (Edgar et al., 2023). This section examines water pollution from the perspective of stormwater runoff and industrial waste.

Concrete & Runoff

Where does this stormwater come from? For the sake of urban development, much of the land's trees and vegetation were replaced by concrete. Cities have built roads, highways, parking lots, industrial buildings, and watersheds of concrete, making cities a comfortable place to live and drive around in. Concrete has a low permeable surface, meaning it has low porosity and high imperviousness, preventing water from being absorbed into the ground. After a rainfall in an urban area, virtually none of the rainwater infiltrates the ground, and it becomes stormwater. From the streets, buildings, and yards, the stormwater travels into storm drains, pipes, waterways, streams, creeks and eventually reaches the ocean (Figure 2.5). This phenomenon results in urban runoff that carries pollutants from the cities into coastal waters through nonpoint sources (Stenstrom, 1999). On the other hand, when there is rainfall in underdeveloped lands with high pervious areas, rainwater infiltrates the soil, where the contaminated water can be degraded by the microbial communities in the soil, decreasing the amount of contaminated runoff (Ewere, 2024).

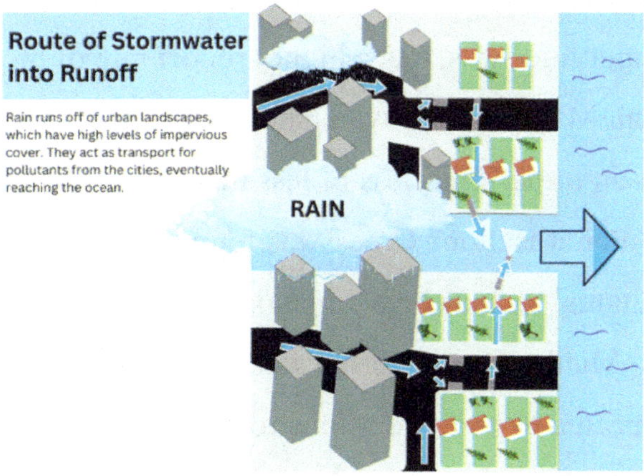

Figure 2.5 *The pathway of rain on an urban landscape, traveling through pavements, streets, waterways, rivers, and ultimately the ocean.*

The relationship between imperviousness, runoff, and water pollution can be best illustrated by observing the runoffs of urbanized watersheds. For example, the watershed in Los Penasquitos Creek in San Diego County, California, went through significant urban development from 1966-2000. Between 1966-2000, the amount of urbanized land increased from 9% to 37% while the total water runoff increased by 4% each year, even though there was no significant change in the amount of rainfall. In total, there had been over 200% increase in runoff from 1973-2000, while flood severity increased by more than 500% from 1965-2000 (White & Greer, 2005). Another example is in Santa Monica Bay, where UCLA researchers observed the runoff characteristics between two watersheds. Ballona Creek watershed, which is mainly built of concrete with high imperviousness, has a higher runoff rate of more than 60% than Malibu Creek watershed, which is less developed and has an imperviousness of less than 30%. Moreover, the stormwater at Ballona Creek was found to have high levels of pollutants, such as heavy metals, that are toxic to marine species. Malibu Creek stormwater, on the other hand, was not suspected to be toxic (Stenstrom, 1999). Overall, it is safe to conclude that city infrastructures with concrete in replacement of soil and vegetation are a primary cause of urban runoffs, transporting pollutants from cities to the ocean (Figure 2.6).

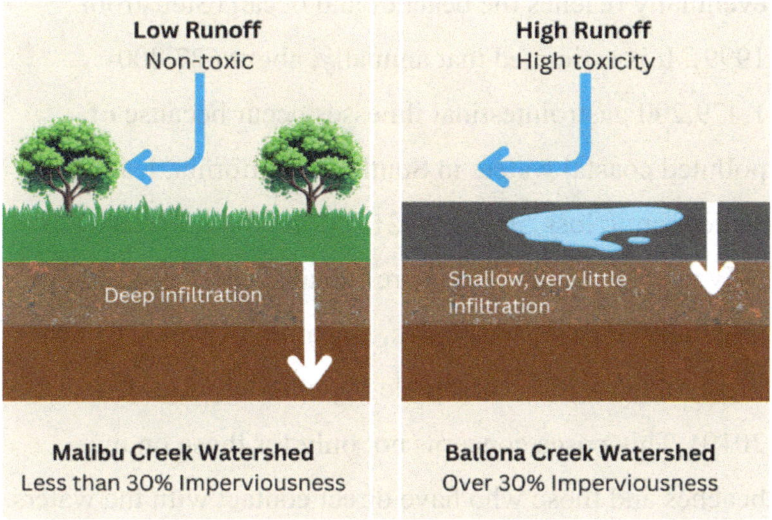

Figure 2.6 Comparison between the rate and quality of runoff between Malibu Creek watershed and Ballona Creek watershed in Santa Monica Bay

Impact on Human Health and Environment

Realistically speaking, how much environmental impact do stormwater runoffs have? According to the study conducted in Santa Monica Bay, it is highly likely that a majority of pollutants entering the ocean in California come mainly from stormwater rather than treated wastewater. We can assume the same for other coastal areas with highly urban cities. An even bigger problem is presented when the stormwater in cities leaks into sewers during a massive rainfall. The surge of stormwater can cause the sewer to overflow, releasing a mixture of sewage and stormwater containing bacteria that poses serious health risks when it eventually reaches the beaches and ocean (Stenstrom, 1999). It is estimated that annually, about 627,800-1,479,200 gastrointestinal illnesses occur because of polluted coastal waters in Southern California, leading to an economic loss of at least 21 million dollars every year (Given et al., 2006). Exposure can also cause ear and eye infections and, in some cases, Hepatitis A, which can be life threatening for vulnerable populations (Aguilera et al., 2019). This poses concerns not only for those on the beaches and those who have direct contact with the waters but also for those on land. When bubbles burst in the ocean, sea spray aerosol is emitted into the atmosphere. This transfers pathogens, chemical compounds, pesticides, etc. into the air, affecting populations on the land via airborne transport (Pendergraft et al., 2023). Pollution from runoff not only affects human lives but also the wildlife habitat in the ocean by altering the salinity, pH levels, and water quality of the ocean. Sewage and heavy metals not only cause defects in many marine species' physiology, but they dramatically increase the alkalinity of the ocean, which many fish populations cannot live in. Feeding off of excess organic waste in the waters, plankton blooms can lead to "red tide" that kills many fish living in the area by depleting oxygen levels (Zutshi & Prasad, 2008).

Industrial Waste & its Consequences

In addition to urban runoffs, another major contributor to water pollution is improper disposal of industrial waste in the environment. Dumping of chemicals, heavy metals of lead, mercury, arsenic, chromium, etc., into the environment and water has unfortunately become common not just in America but around the world. It is estimated that in 2022, 3.4 billion pounds of chemical waste was released into the environment by facilities. These chemicals were listed under the Toxic Inventory (TRI), which are known to be carcinogenic and cause other chronic health effects in humans (US EPA, 2024). In the U.S. alone, there have been many lawsuits involving large corporations dumping industrial waste into the environment, contaminating water sources, and harming the surrounding community (Gaber et al., 2023; Konwinski & Olayinka, 2022; Schmidt et al., 2024; Yeager, 1993). Once investigations started, further discoveries were made about the hazardous effects. In Ohio, many animals were found dead after ingesting chemicals that had leached into a river near the dumping site. In humans, studies observed that the workers at these corporations and people living in the surrounding community who were exposed to these chemicals/wastes had higher rates of cancer, increased cholesterol, thyroid diseases, and hypertension (Anderson-Mahoney, 2008; Konwinski & Olayinka, 2022).

This poses the question: to what extent were the corporations aware of the hazardous effects of their practices, and since when? According to a study led by Nadia Gaber that investigated one such corporation, they already knew the toxicity effects at least a few decades before public findings. The study examined and compared the internal documents to published scientific literature and found that although there were internal reports of workers suffering health issues at the plants and animal studies that pointed towards the toxicity of the chemicals, they were not made public nor published. This indicates that these companies have been intentionally suppressing unfavorable research findings, ignoring future implications, and continuing to dump hazardous chemical waste into the waters for the sake of maximizing profits (Gaber et al., 2023). Yet, many of those same companies that are behind dumping industrial waste are the same ones who acquired technology to treat contaminated waters and built treatment plants (DuPont Media, 2020; US EPA). It became inevitable that the companies would be paid to remove, filter, and purify the toxic chemicals from the water. In the end, those companies managed to generate profits from not only contaminating the waters, but for cleaning them up as well.

Projection

In nearly 30 years from now, it is projected that the global human population will increase by almost 2 billion (United Nations, n.d.), and 70% will live in large urban areas (Parsons, 2021). This will naturally increase the demands for concrete and urban structures, and higher costs for waste management. In turn, this can accelerate the decline in water quality and compromise the health of human beings and the well-being of our Earth. Chemicals such as PFAS (Per- and polyfluoroalkyl substances) are not biodegradable, they will remain forever in the environment, hence they are known as forever chemicals (Konwinski & Olayinka, 2022). Even though there are treatment methods being developed, PFAS has already infiltrated many parts of our environment, including our bodies. Already, it is estimated at least 45% of tap water in America has been contaminated with PFAS (Smalling et al., 2023) and is found in the bloodstream of over 98% of the US population

(Shearer et al., 2020). As PFAS cannot be degraded by biological processes, they will remain in our bodies, accumulating over time. Over 8 million new chemicals are synthesized each year (Escher et al., 2021) and only a few are tested for toxicity (Muir et al., 2023). Even if they are, we will not know its long-term effects until a long time after it is already introduced to our environment. By that time, many lives would have been affected.

What can we do? Although civilization has achieved remarkable technological advancements in the 20th and 21st centuries, we have harmed the environment for the sake of our convenience and profit. For the sake of convenience and comfortable living, we built cities made of concrete. For the sake of profit, corporations ignored the warning signs and continued their industrial waste dumping. Now, the damage humanity has brought to the environment is being reflected back to us. To ensure the prosperity and well-being of humankind, it requires the health and environmental integrity of Mother Nature. If we can heal and restore the natural environment, then surely humanity can see positive changes in turn. A fundamental shift in our collective perspective—one that recognizes the intrinsic value of all living things—is essential. Only when we acknowledge the interconnectedness of life on Earth can we develop more sustainable and responsible practices to address the issue of water pollution.

Section 3

Soil Science and Agronomy in Relation to Climate Change

Written by: Jinka Kawasaki

Soil is fundamental to all terrestrial life, providing essential ecosystem services, including net primary production, food and fiber production, water storage and quality, and biodiversity conservation (Lal, 2010a) (Figure 2.7). Healthy, well-managed soils have historically supported great civilizations, including those of Mesopotamia, the Indus Valley, the Maya, and the Aztecs. Soil is the foundation for the food we grow, the livestock we care for, and the home of various organisms. As a living entity, soil supports approximately 8.7 million species and maintains a dynamic equilibrium with its environment. However, all of these services are dependent on the quality of the soil. Soil quality is defined by its capacity to support ecosystem services, which depends on the quantity and quality of the soil organic carbon (SOC) pool within its root zone (Lal, 2010a). The role of soil organic carbon (SOC) is in regulating terrestrial ecosystem functions, supplying diverse energy sources for soil microorganisms, maintaining soil structure, and regulating the availability of organically bound nutrients. However, soil resources are finite, non-renewable on a human timescale, and highly vulnerable to disturbances from both natural forces and human activities such as deforestation, drainage, and tillage. When this delicate balance is disrupted—whether by accelerated erosion from water and wind, salinization, depletion of soil organic matter (SOM) and nutrients, water-air imbalances causing drought or inundation, elemental imbalances leading to nutrient deficiencies or toxicities, or a decline in soil biodiversity—degradation occurs.

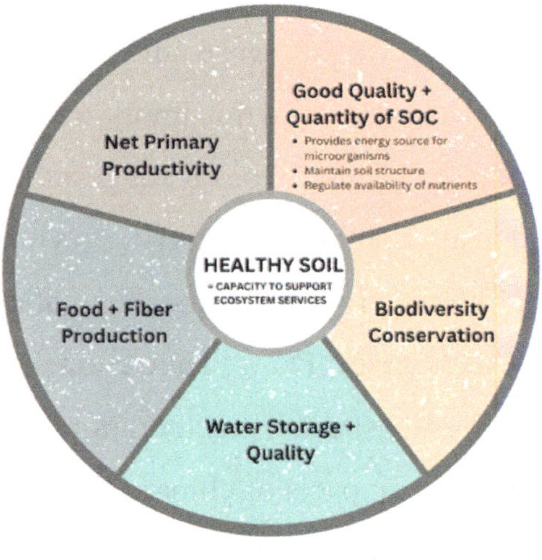

Figure 2.7 Valuable ecosystem services healthy soil provides to the environment.

Currently, 40% of Earth's surface is used for agriculture, of which 70% of freshwater is used for irrigation. That's almost half of the Earth's surface and more than half of Earth's freshwater being used for agriculture. Despite a shocking amount of our resources invested into agriculture, the reality is that 2 in 7 (~1.2 billion in 2022) are malnourished. 1 in 4 (2 billion people) are under-nourished. More than 3 billion people do not have access to safe or healthy food. In addition, ⅓ of Earth's soil that the agriculture industry depends on is degraded. Soil degradation, characterized by a reduced ability to support functions and deliver ecosystem services, results from accelerated erosion, salinization, nutrient imbalances, acidification, depletion of (SOC), loss of biodiversity, and deterioration of soil structure and tilth (Lal, 2023). Although an immense amount of resources are invested into agriculture, does this imply we lack the resources to feed everyone? Does it suggest that our current resources are insufficient to sustain the growing population? With the growing human population, another important question arises: Do we have enough food to sustain this increase, especially as food production is closely linked to the impacts of climate change? This is a highly interconnected and complex challenge requiring an interdisciplinary approach. The objective of this topic is to highlight the positive feedback loop and the relationship between climate change, agriculture, soil health, and human suffering. Recognizing this destructive feedback loop emphasizes the need for a new approach - one that goes beyond technological advancements and incorporates the internal attitude of environmental stewardship.

Effect of Climate Change on Agriculture and Soil Degradation

Projected climate change can trigger extreme climatic events and increase the risks of droughts, floods, water scarcity, and storms (Beddington, 2012). This requires adaptation, as these effects will vary across different regions. In Sub-Saharan Africa and Asia, food security in many densely populated regions is threatened by the uncertainties and challenges intensified by climate change. Rising sea levels could impact Bangladesh's highly productive agricultural areas (Lal et al., 2011). Food security remains a concern in China, the world's most populous nation (Brown, 1995), with climate change expected to further exacerbate the issue (Tao et al., 2009). Due to changing environmental conditions due to climate change, Inuit food systems face increased vulnerability (Ford, 2009). Understanding climate change, even in developed nations, such as the U.K., is important for managing food demand and security (Bows et al., 2012). In addition, crop yields are dependent on various aspects of

Earth's shifting climate, including rising temperatures, altered precipitation and cloud patterns, elevated CO_2 levels, and more frequent extreme weather events such as heat waves, floods, and droughts (Kerr et al., 2022). The Intergovernmental Panel on Climate Change (IPCC) 6th assessment report found crop yields across most regions and food groups are harmed due to the results of climate change (IPCC, 2023). Worldwide, agriculture depends on weather and climate conditions, making it vulnerable to the changing impacts of climate change.

Additionally, climate change contributes to soil degradation through the processes of desertification-erosion, and salinization. With 41% of the continental area classified as dry or arid, these regions are highly vulnerable to desertification—a form of land degradation that diminishes the quality and functionality of soil, vegetation, water, biodiversity, and climate. Climate change accelerates desertification by altering factors such as solar radiation, temperature, rainfall, and extreme weather events. Extreme weather events also include droughts, which could be a long-term decline of precipitation, prolonged periods of decline in surface runoff, depletion of the aquifer, and depletion of the soil's available water capacity. Drought and soil degradation are closely interconnected, both worsening due to projected climate change and further intensified by positive feedback loops linked to human activities (Figure 2.8). Soil erosion, which is the physical removal of soil, in its severity, causes loss of effective rooting depth, reduction in plant-available water capacity, depletion of soil organic carbon, decline in soil structure, and decline in soil quality. Areas experiencing frequent and intense weather events, such as higher rainfall, stronger winds, and increased runoff, have enhanced the sediment-carrying capacity. Especially as soil organic carbon concentrations decline, soil erodibility rises due to reduced aggregation and weaker aggregate strength (Lal, 2012).

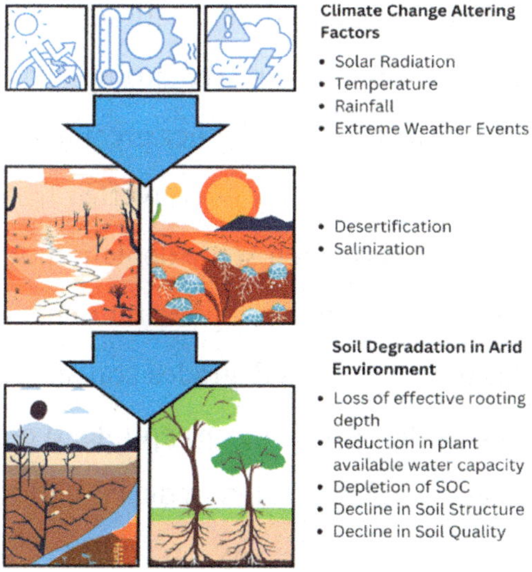

Figure 2.8 The impact of climate change in the degradation of soil in arid environments.

Soil degradation resulting from climate change and agricultural practices ultimately cycles back to negatively impact agriculture itself. Soil resources are limited and susceptible to degradation from land misuse and poor management. Soil degradation diminishes agricultural productivity (Lal, 2013a). Agroecosystem soils have lost 25–75% of their original soil organic carbon (SOC) pool due to land misuse and poor soil management. The extent of this loss is greater in soils subjected to extractive farming compared to science-based agriculture, in coarse-textured soils rather than fine-textured ones, and in tropical climates more than temperate regions. Since the advent of settled agriculture 10–12 millennia ago, land use and farming practices—such as deforestation, biomass burning, wetland drainage, plowing, and manuring—have been significant sources of CO_2 and other greenhouse gases (Ruddiman, 2003). When land is converted from natural ecosystems to agroecosystems, the depletion of the soil organic carbon (SOC) pool and plant nutrients due to extractive farming practices degrades soil structure and tilth. This makes the soil more susceptible to crusting, compaction, accelerated erosion, reduced biological activity, loss of species diversity, and other forms of degradation (Lal, 2013a). The negative effects of soil degradation on agronomic productivity can escalate exponentially as its severity increases from mild to severe (Lal, 2013a). Agricultural soils contain 25% to 75% less SOC than their counterparts in undisturbed or natural ecosystems (Lal, 2004). The soil organic carbon (SOC) stocks in soils from managed ecosystems, such as in agriculture, urban, mining, and recreational land, are diminished. When these natural ecosystems are converted to agricultural land, the SOC levels are depleted due to lower biomass carbon inputs such as roots or litter falls and due to erosion and leaching of dissolved organic carbon. Soil organic carbon is lower in density and concentrated near the surface, which makes it susceptible to erosion from water and wind and accelerated depletion. In agriculture, additional factors such as overgrazing (Dlamini et al., 2016) and plowing (Lal, 1976) further contribute to SOC depletion. The depletion of SOC, along with other forms of soil degradation, increases land scarcity and intensifies food insecurity (Gomiero, 2016).

Agricultural practices directly influence soil health and degradation, which in turn impacts agricultural productivity, as soil health is a critical determinant of crop performance.

Effect of Agriculture and Soil Degradation on Climate Change

In turn, soil degradation and unsustainable agriculture further contribute to climate change. Soil worldwide serves as a major reservoir of reactive carbon, playing a crucial role in regulating the global carbon cycle, atmospheric composition, radioactive forcing, and ecosystem services. Consequently, soil carbon sequestration is essential for mitigating climate change and limiting global warming to 2°C (Lal, 2013b). 2 of the 5 major global C pools are pedologic and biotic, combined is the terrestrial C pool. Together, it equates to 3120 Pg, which is four times the atmospheric pool. Depending on land use and management practices, the terrestrial carbon pool can act as either a sink or source of atmospheric CO2. Soil degradation and desertification further accelerate carbon emissions from terrestrial ecosystems (Lal, 2010b). Agriculture impacts the climate by emitting greenhouse gases (GHGs) linked to agricultural activities, such as livestock production and nitrogen fertilizer use, as well as through the expansion of farmland into natural vegetation. Conversely, climate change affects crop growth directly by shifting temperature and precipitation patterns and indirectly by influencing factors such as weed competition and pest outbreaks (Bajželj, 2014).

Additional Component in the Feedback Loop

Many scientists highlight an additional dimension in the relationship between climate change, agriculture, and soil health: human suffering. This suffering not only results from these challenges but also contributes to the reinforcing feedback loop.

Dr. Rattan Lal, a renowned soil scientist, mentions in *Climate Change and Soil Degradation Mitigation by Sustainable Management of Soils and Other Natural Resources,* "A prude management of the anthropocene through an earnest attempt by man to domesticate himself can reverse the downward spiral involving mutual reinforcing and positive feedback. The severity of the problem is increased manyfold because soil degradation would increase risks of climate change and vice versa in an accelerating spiral."

Therefore, environmental security and human security are closely interconnected. The driving force of environmental degradation and climate change is human-induced disruption of the fragile balance in ecologically sensitive eco-regions. There is a strong connection between poverty, desperation, societal collapse, and environmental decline. The gap between agronomic production and food demand, already affecting more than one billion people, is likely to increase by the projected climate change. Soil degradation increases the risk of climate change and vice versa, creating a positive feedback loop. The positive feedback loop between land use conversion, climate change, and desertification/erosion, driven by feedback and thresholds, cannot be ignored. Feedbacks and thresholds amplify the risks, causing degradation to increase disproportionately faster than population-induced land-use conversion (Lal, 2012).

Risks of soil degradation are intensified by increasing frequency and severity of extreme events and have consequences for civil stability and security; declining farm income and rising unemployment contribute to social unrest and economic instability. Political instability, civil strife, extremism, and violence often emerge in the poorest and most deprived regions, particularly in Africa (Annan, 2016), where soil degradation is widespread, agricultural productivity is low, and hunger and malnutrition are severe. The depletion of the soil organic carbon (SOC) pool, soil degradation, recurring droughts, inefficient use of fertilizers and other inputs, poor crop yields, persistent poverty, and high infant mortality due to malnutrition pose significant threats to global peace and security. The interconnected health of soil, plants, animals, people, and ecosystems underscores the urgency of addressing soil degradation. Furthermore, its impact on human health is profound (Kemper and Lal, 2017).

Various factors, including human activities and climate change, continue to accelerate soil degradation. Given that soil quality relies on key components such as soil organic carbon (SOC) and soil organic matter (SOM), their preservation and maintenance are crucial for long-term agricultural sustainability.

This merely scratches the surface of the interconnected challenges at play. In conclusion, we do have the necessary resources to feed the growing population, but until we can manage it in a sustainable way, we will not. We have many contributing factors, such as anthropogenic influences and climate change exacerbating soil degradation. As soil quality is dependent on many factors, such as SOC and SOM content, this needs to be maintained.

Soil health, plant health, animal health, human health, environment health, and planet health are all one and indivisible (Lal, 2024). Soil stewardship and care should be in every fruit and vegetable we eat, every grain milled into bread, every drop of water we use, and every breath we take (Lal, 2019). That is the depth of care that should be ingrained in our relationship with the natural world. Therefore, a "business as usual mindset" is unacceptable.

Chapter 3

An Innovative Solution

Written by: Maiko Shimogawara and Yuta Horii

Written by Maiko Shimogawara & Yuta Horii

1. Offering the Key to the Solution

When people observe untouched nature, they are immersed in its beauty. This beauty arises from the hidden order within the natural world. Humankind must learn how to follow that order rather than interfere with the natural world. To do this, a fundamental transition must take place. The self-centered approach that humankind has long imposed upon nature must shift towards a practice of stewardship rooted in love. As discussed briefly in Chapter 2, the solution lies in understanding how humans stand within this higher order of all things. How can this shift occur, moving from past behaviors to a stewardship based on love? What does it mean to practice stewardship in this way? What is the higher order, and how do human beings fit within it? These are critical questions that will be explored in detail.

However, to arrive at this new understanding, it is essential to examine the cause of these current problems. Chapter 2 outlines how human behaviors have caused harm to the Earth, which in turn has negatively impacted human lives. Despite being driven by the motivation to improve the quality of life for all, the current situation presents several pressing environmental problems that many scholars feel must be addressed now. However, the fundamental solution is not about addressing each of these problems individually and finding specific solutions for them.

Chapter 2 highlights a few of the environmental problems, yet the reality is that many more remain unmentioned. Each of these are not the fundamental problem, but rather they are symptoms. They are an expression of a deeper unresolved issue. Therefore, the new solution must tackle the root cause of humanity's damaging relationship with nature.

1.1 Mind Structure and Behavior

To arrive at that root cause, the inseparable link between the human mind and its behavior must be explored. Humanity has been driven by a desire to seek for the hidden laws of nature and advance science and technology to create a pleasant living environment. "From the commonplace affairs of individuals to the great events that shape the course of history, each is at root an expression of the human aspiration for ever greater happiness and joy" (Moon & Eu, 2006, p. 1). The development of science and technology is no exception.

Simply put, as an expression for the human's pursuit of joy and happiness, they have sought to develop science and technology to improve their living environment.

If the structure of the mind and behavior were to be defined, it would be thus:

- **Heart**: The heart is the core of the human mind and the emotional impulse to seek joy through love
- **Mind Structure**: The mind consists of three faculties: emotion, intellect, and will, with the heart at the center of all three. In the mind, we first establish a purpose, and then, as our heart constantly stimulates the intellectual, emotional, and volitional faculties, we create a design or plan to implement that purpose. This structure of the mind also shapes and determines human consciousness
- **Behavior**: Based on the design or plan in the mind, we engage in emotional, intellectual, and volitional activities through the body. These activities involve interacting with or controlling material objects (such as things in nature and manufactured goods) and human partners

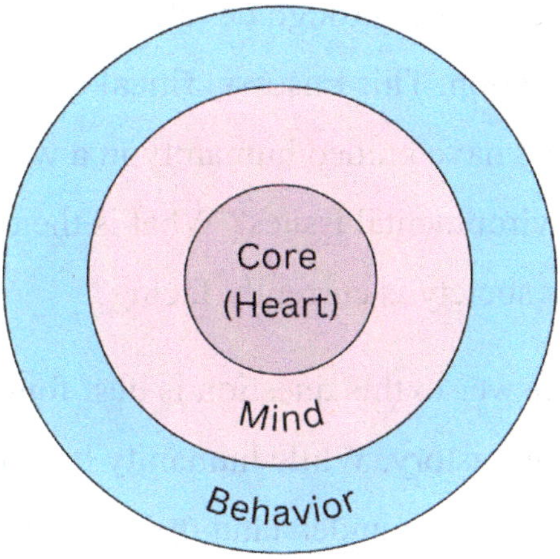

Figure 4.1 *Inner Sungsang and Inner Hyungsang centered on Heart.*

Adapted from *New Essentials of Unification Thought* by S. Lee, 2014, Unification Thought Institute (p. 25). Copyright 2014 by Unification Thought Institute.

Development of Science and Technology in Context of Mind Structure and Behavior

The desire to feel joy by creating a better living environment sets the purpose for human beings to understand nature to advance science and technology. However, has that expression of that heart translated well into the mind structure and behavior? The reality is that humanity has a long-standing tendency to engage in actions that have harmed the environment and the natural world. Considering how human behavior has led to many of the current environmental crises, it can be inferred that something went amiss with the mind structure. In other words, the design that humankind constructed to experience that joy and happiness through development transpired in the wrong direction. This raises a critical question: What mind structures have guided humanity in a way that has led to modern environmental issues? What is the cause behind the effect that society is currently facing?

The answer to this question is best found in the study of human history. While humanity has pursued its fundamental calling to understand the mechanisms of nature and advance science and technology to create a pleasant living environment, these advancements have also been used to justify exploitation and domination. Societies have often been driven by a need to exploit and extract resources from colonized territories. This trend can be observed even from the primitive years of human history.

Humanity started as a hunting and gathering species where they spent much of their day going out into nature to find something to eat for the group. However, this is a rather inefficient method of promoting the growth of a population. Hunting and gathering require a lot of energy, and the source of food is often unreliable. Eventually, due to certain groups living in environments suitable for agriculture, some were able to transition into agriculture. As a result, they were able to have a reliable source of food and create specialized jobs. These advancements allowed those specific groups to experience a population growth that neighboring hunting and gathering groups could not (Diamond, 1998).

What ensued was that these agricultural groups then started claiming territories over the other groups that were still stuck in the primitive method of hunting/gathering. Over the span of time, this led to the development of nation-states as well as empires. Concurrently, this process also led to advancements in other fields, such as medicine and transportation.

An even stronger example of the application of this belief structure can be seen in the time of the industrial revolution that began in Great Britain in the 18th century. The world experienced massive population growth beginning with the revolution due to the increased acquisition of natural resources.

The exponential growth in population was further supplemented by advancements in other fields, such as medicine and a stable food supply. As the name suggests, the Industrial Revolution marked a major shift from an agrarian economy to an industrial society.

However, rooted in this development was the need to facilitate colonial expansion over indigenous populations. Development in other fields, such as medicine, transportation, communication, military, and technology, were all used to create an exploitative dynamic, with European powers largely dictating the direction of scientific research in colonized regions to benefit their own interests. Remarkably, the industrial revolution appears to be a significant starting point for widespread environmental problems. The increase in the use of coals and deforestation in the name of colonial conquest caused issues of air and water pollution that persist today.

The core principle behind the development of science and technology is not inherently bad. It was rooted in the desire to feel joy by creating a pleasant living environment. However, the pursuit of this happiness was misdirected. Within humanity's mind structure, they adopted a design that focused on pursuing happiness through expanding territory and using natural and human resources for their own benefit. As a result, even though science and technology did advance, the way human beings utilized and sought after it came to resemble self-centered domination. Consequently, humanity's behavior has become harmful to the Earth, risking the wellbeing of humanity as well.

Thus, the fundamental cause of environmental issues is the current structure of the mind. In other words, it is the human consciousness. The current human consciousness has been the fundamental cause of the harmful behaviors against the Earth. This has led to the interference against the natural world, damaging the Earth and ultimately damaging humankind themselves.

1.2 Consciousness in the Universe and the Holomovement

To address the issues within human consciousness, this section will explore whether consciousness in the universe, or the higher order of all things, exists. If such a consciousness exists, it could serve as a model for transforming human consciousness. By examining how individuals relate to this concept, corrective efforts can be initiated, leading to effective measures against the environmental crisis.

The study of ecology explains that the universe and all things share an interconnected relationship. This holds true for all living and even non-living beings. Observations such as the carbon, water, and solar cycles demonstrate the interconnectedness of all things. There are no variables that stand independent of the rest; rather, they all share a circular relationship that is continuous and interrelated.

What exactly allows this interconnectedness and cycle to take place in the natural world? The answer to this may be best addressed by the concept of Holomovement. Proposed by Dr. David Bohm, his theory of Holomovement describes the universe as an unbroken wholeness of the totality of existence – an undivided, flowing movement without borders. The Holomovement, as explained by Bohm, was "a holistic understanding of our universe as a manifestation of the consciousness of our universe" (Bohm, as cited by Kuntzelman & Robinson, n.d.). Traditionally, in classical physics, reality was understood to consist of matter and energy. However, Dr. Bohm suggested that reality consists of matter, energy, and consciousness. Although this thought was not widely accepted at the time, the experimental tests of Dr. John Clauser, Dr. Alain Aspect, and Dr. Anton Zeilinger proved a phenomenon called Quantum Entanglement, winning the Nobel Prize in Physics in 2022. This significantly increased the plausibility of a possible interconnectedness of all things. Briefly explained, Quantum Entanglement is a phenomenon where two particles in completely different locations behave like a single unit. In other words, if a change occurs in one, it will cause a change in the other instantaneously, bypassing space and time. The measurement of the state of one particle determines the state of the other, indicating a very strong correlation (Nobel Prize, 2022).

"Quantum entanglement shows that the two particles are not independent of each other, but rather that their properties are strongly correlated in a way that classical physics cannot explain. This phenomenon defies the classical views that objects should be described independently of each other. It suggests that on the quantum level, there is an intrinsic connection between particles that transcends the boundaries of time and space."

(Zeilinger, 2010)

This nonlocal connection may be seen as the continuous work in the universe. In other words, the work of an internal nature or consciousness is causing a change in the external form (the state of particles).

2. An Innovative Approach

Nature is self-sufficient and capable of sustaining itself through intricate cycles and interconnected systems. Therefore, if there is an interconnectedness in the universe as Dr. Bohm saw it in the form of Holomovement – guided by a consciousness that creates the universe and unfolds all the mathematical laws and physical phenomena – then the environmental issues caused by human beings can be seen as a result of human consciousness disrupting this Holomovement.

Thus, in this solution- oriented approach, the first step is to investigate whether a consciousness operates within the universe to maintain the interconnectedness among all beings. If such a consciousness exists, the next step is to explore its nature and role. The inquiry then shifts to understanding why only one type of being – human beings – possesses the capacity to disrupt this natural interconnectedness, thereby causing environmental harm.

If human consciousness is indeed the root cause of environmental issues, the logical progression is to examine how a collective shift in consciousness can be achieved.

This shift should focus on altering behavior to restore a harmonious relationship between humanity and the natural world. Ultimately, this process will not only address environmental challenges but also realign human existence with the interconnectedness that sustains life on Earth.

2.1 Consciousness in the Universe

The Conscious Energy Guiding Growth and Development in All Beings

As mentioned above, the first step is to examine whether there is a consciousness in all beings, with a focus on the motion of growth and development. Living beings are endowed with life, which refers to conscious energy – consciousness combined with energy – latent within them. This conscious energy guides the growth of living beings, as it is the unity of consciousness and energy that directs the motion of life. The movement of conscious energy is, in essence, the movement of life itself.

Conscious energy is characterized by autonomy and governance of principle. Autonomy refers to the ability of living beings to direct their own motion without influence from other beings. For instance, while the Earth revolves mechanically, living beings are able to control their own growth and respond to their environment. This can be seen in how a seed eventually grows into a plant. Once planted, the seed sprouts, develops a stem, and eventually produces leaves. This force of growth is the action of the autonomy of the principle. On the other hand, "dominion of the principle" refers to the ability to exert influence over one's surroundings. While the autonomous character of conscious energy allows the seed to grow into a plant, concurrently, the plant has an influence on its surroundings. It provides animals with oxygen and attracts bees and butterflies by blooming. This aspect is the dominion of the principle (Lee, 2014).

Dr. Bohm stated, "According to the implicate order, the seed is continually providing inanimate matter in the environment with new information that leads it to produce the living plant or animal. Who is to say, then, that life is not immanent, even before the seed is planted?"

(Wilber, 1982, p. 193)

Autonomy and governance of principle suggests one key point. The growth of living beings, due to the life force inherent within them, follows a developmental motion. All created beings are endowed with the purpose of existence, meaning that the life force within them is conscious of this purpose. Therefore, the growth of living beings is a movement towards a goal from the very beginning, with a definite direction determined by their inner life force, which is their conscious energy.

For example, the conscious energy within the seed of a plant causes it to grow toward the goal of becoming a tree that bears fruit. Similarly, the life within a fertilized egg drives it to develop into an adult animal. This sense of purpose is intrinsic to all living beings.

In fact, the development of the universe also shares this characteristic. According to the Big Bang Theory, the universe began as a mass of energy, and through a great explosion, it began to expand. Over time, galaxies formed, stars emerged, and planets were created, including the Earth. Life eventually appeared on Earth, and humans emerged. The question then arises: Is the development of the universe any different from the growth of living beings? Is the development following a certain direction, or is it simply just a consequence of the physiochemical laws at work?

At a short time scale, the development of the universe may appear to follow only physicochemical laws. However, when viewed over a much longer period of time

– say, several billion years – it can be discerned that while certainly governed by the physicochemical laws, the universe is developing in a definite direction, much like the development of living beings.

This can be explained further with Cosmic Inflation. Cosmic Inflation Theory suggests that the universe experienced an incredibly rapid expansion in the first fraction of a second (from about 10^{-36} to 10^{-32} seconds) after the Big Bang. During this inflationary period, space expanded exponentially and smoothed out the universe. This set the stage for the formation of galaxies, stars, and all cosmic structures that can be seen today (Guth, 1981). Over time, this expansion slowed down, but the universe continued to evolve in ways governed by fundamental forces and laws of physics (Linde, 1982).

In cosmic inflation, the rapid expansion of the universe set the framework for space and time, creating conditions that allowed matter to form and evolve. As space expanded, energy and matter interacted under the influence of gravity and other forces (Guth, 1981). It follows certain laws of physics that guide the formation and interaction of particles, galaxies, and elements, much like how living beings evolve toward greater complexity (Rees, 1999).

Thus, the long-term development of the universe can be viewed as a process where the universe's laws and the forces of nature guide a continual exchange of energy and matter, which, over billions of years, shapes the universe into a more complex and ordered system. This might not only be an evolution of matter but could also be seen as a broader, cosmic progression toward increasing complexity, echoing the idea of development over time that is observable in biological organisms.

Therefore, the universe itself may be undergoing a process of development driven by conscious energy in the same way that living beings are guided by conscious energy within them. What has given this kind of direction to the development of the universe is the consciousness latent within the universe. This can be called "cosmic consciousness".

Carroll suggested "the arrow of time," which indicates that time flows in one direction, the past to the future, with relation to the laws of physics. And the role of time in the evolution of the cosmos owes its condition before the Big Bang (Carroll, 2010).

This demonstrated that consciousness exists in all beings, as the life force within living beings guides their growth with purpose. Moreover, the universe itself, though governed by physical laws, unfolds in a conscious direction driven by a latent cosmic consciousness.

Cosmic Consciousness and Protoconsciousness

Unification Thought defines the relationship between cosmic consciousness and the consciousness in a being as **protoconsciousness**. Protoconsciousness refers to the original cosmic consciousness that enters cells or tissues, becoming individualized as life. It is described as a lower level function of the cosmic mind. Protoconsciousness is life itself and encompasses the following qualities,

Sensitivity: the ability to perceive information intuitively

Perceptiveness: the ability to maintain the state of perception

Purposiveness: the willpower to actualize a purpose

In other words, protoconsciousness is the subconsciousness present within the cells and tissues of living beings, serving as the life force that enables growth and development. So, it can be said that living beings become alive once cosmic consciousness has entered their cells, tissues, and organs.

Functions of Protoconsciousness

Now the function of protoconsciousness will be explained. Protoconsciousness has various functions, including the reading of the genetic information (code), acting according to the direction of the information, and transmitting the information.

1. Reading the Genetic code: When cosmic consciousness enters a cell, protoconsciousness first reads the genetic information contained in the DNA of that cell. This genetic code provides the instructions for the cell's function, growth, and behavior

2. Acting according to genetic instructions: After reading the genetic code, protoconsciousness causes the cells and tissues to act in affordance specified in the genetic information. These actions include the development and growth of cells, tissues, and organs. It also governs the relationship that these cells, tissues, and organs form with one another

3. Transmission of information: Protoconsciousness plays a key role in the transmission of information. The information within cells and tissue is sent to the central nervous system through peripheral nerves.

> The central nervous system, in turn, sends directions back to the cells and tissues via centrifugal nerves. This back-and-forth exchange of information is mediated by protoconsciousness

These functions are all made possible by the protoconsciousness' qualities of sensitivity, perceptivity, and purposiveness. These qualities ensure that the protoconsciousness effectively guides and coordinates the development and functioning of living organisms. As protoconsciousness carries out its functions over time, it facilitates the development of the protoimage – the initial, underlying form of an organism or system – as well as the clarification of relationships among the various components of the organism. This process includes the gradual formation and refinement of the organism's structures and interaction.

In summary, protoconsciousness refers to the original cosmic consciousness that enters cells or tissues and acts as their consciousness. It possesses three key faculties: sensitivity, perceptivity, and purposiveness, which correspond to the human mind's faculties of emotion, intellect, and will. As a fundamental force, it reads and implements genetic information, facilitates communication within the body, and ensures the development and growth of the organism according to the genetic plan. This holds true even for minerals as well. However, in minerals, it surfaces only as physiochemical functions.

Ultimately the main point is that cosmic consciousness, which is latent within the universe, bestows consciousness upon each being as protoconsciousness, guiding them with a clear purpose and direction for growth and development. In this way, all beings are interconnected through the influence of cosmic consciousness.

According to Dr. Bohm, "If the immanence is pursued more and more deeply in matter, I believe we may eventually reach the stream which we also experience as mind, so that mind and matter fuse."

(Wilber, 1985)

2.2 The Purpose and the Hidden Order in the Universe

As explained, the protoconsciousness enters cells or tissues and becomes individualized in life, acting as their consciousness. According to the complexities of each being, there is more. There is the inner quality, though invisible, that possesses a certain structure, which is manifested visibly in the particular outer form. Unification Thought calls this Internal Nature. The outer form, which is also another aspect of all things, is called their external form. All beings exist within the reciprocal relationship between these dual characteristics – internal nature and external form. In this section, internal nature and external form will be explained more at various levels of existence in order to uncover the purpose of existence and the order of the universe.

Dual Characteristics Exist in Every Entity

Unification Thought suggests that, although the complexities and qualities may vary, all beings possess the dual characteristics of internal nature and external form. Internal nature refers to the invisible, immaterial aspect of created beings, such as their faculty and nature. External form refers to the visible aspect of created beings, such as mass, structure, and shape. For example, in minerals Internal nature is a physicochemical character, and External form is structure, shape, and so on, composed of atoms and molecules (Lee, 2014).

Plants have their own unique Internal Nature and External Form. The Internal Nature unique to plants is life (autonomy), and the External Form unique to plants is their cells and tissues, which compose their structure and shape—in other words, the body of a plant. The function of life (autonomy) is the ability to grow while maintaining itself as an individual being. While plants possess their own unique Internal Nature and External Form, they also contain the elements of Internal Nature and External Form of the level of minerals. In other words, plants contain mineral matter.

The same holds true for animals as well; in fact, in a higher position. "In animals, there are aspects of Internal Nature and External Form that are unique to animals and so they exist on a level higher than that of plants. The Internal Nature unique to animals is instinct, and the External Form unique to animals is their structure and shape which includes sense organs and nerves. Animals have both mineral matter, which contains the Internal Nature and External Form of the mineral-level, and they also possess plant-level Internal Nature and External Form; all the cells and tissues of animals exist on this level" (Lee, 2014, p. 106).

For human beings, they are composed of the human mind and the human body which are unique to human beings. They serve as the internal nature and external form, respectively. Mineral matter is also contained in the human physical body, and in this sense, the human being has mineral-level internal nature and external form. To go even deeper, the human physical body is also composed of cells and tissues and, therefore, has a plant-level internal nature and external form as well. Like animals, human beings have sense organs and nerves, and hence, the internal nature and external form correspond to animals. The animal-level internal nature of human beings is the instinctive mind. However, on top of all that, the human mind has the faculty of mind to pursue the values of truth, goodness, beauty, and love on top of the instinctive mind.

From this, as the levels of existing beings ascend – from minerals to plants, to animals, and to human beings – The internal nature and external form become more substantial and elaborate layer by layer. In Unification Thought, this is known as the *Layered Structure of Internal Form and External Form in existing beings (Illustrated in Figure 4.2).*

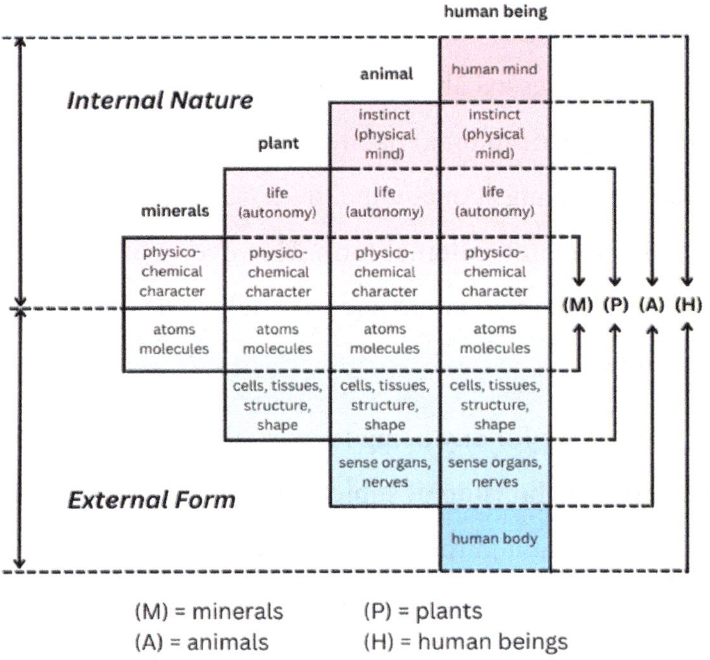

Figure 4.2 Layered Structure of Internal Form and External Form in Existing Beings.

Adapted from *New Essentials of Unification Thought* by S. Lee, 2014, Unification Thought Institute (p. 107). Copyright 2014 by Unification Thought Institute.

Another pair of dual characteristics, Yang and Yin, is discussed in Unification thought as the essence of existence. These dual characteristics represent the attributes of Internal Nature and External Form. For the purposes of this guide, this discussion will be deferred to a later opportunity.

Connected Being of Two Stage Structure of Existence

As discussed above, all beings possess the dual characteristics of internal nature and external form, and higher-level beings contain the lower-level beings of internal nature and external form. However, not all elements of internal nature and external form simply exist within a being.

In order for any being to exist, they must be in a correlative relationship of subject and object, formed around a common purpose, both between two elements within a being (inner) and between one being and another (outer). This leads to an action of giving and receiving a certain element or force. This type of action between subject and object is called *give and receive action.* Through this action, the entities involved maintain their existence and are able to move, change, and develop. Here, the subject is the active, initiating element that exercises dominion and drives action or influence in the relationship, while the object is the passive, receiving element that responds to or is acted upon by the subject, which can include any being or entity within a relationship with the subject.

These two forms of give and receive action – inner

and outer – are necessary for existence and can be understood through the concept of the *four position foundation.* The inner give and receive action forms an inner four position foundation, while the outer give and receive action forms an outer four position foundation. Both are essential for any being to maintain its existence, change, and develop. The give and receive action gives rise to the state of union between the subject and object, or a new being. The formation of the inner and outer four position foundations of any being is known as the *two-stage structure of existence (Figure 4.2).* Without the harmonious and reciprocal inner and outer give and receive actions, no being can truly exist or grow.

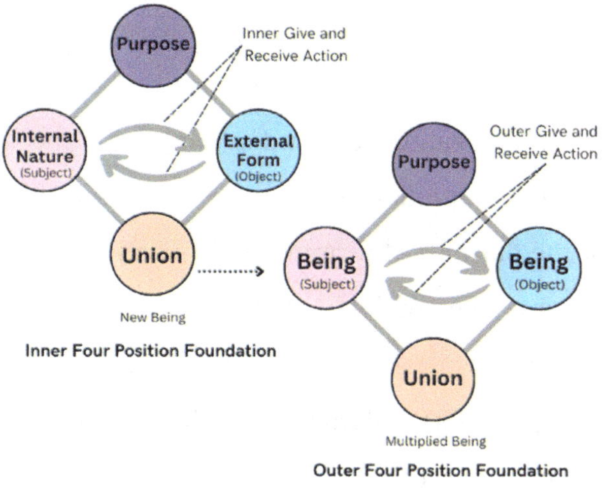

Figure 4.3. Two-Stage Structure of the Original Image.

Adapted from *New Essentials of Unification Thought* by S. Lee, 2014, Unification Thought Institute (p. 51). Copyright 2014 by Unification Thought Institute.

From this structure of existence, it can be concluded that all beings are interdependent and interconnected. The harmonious inner and outer give and receive actions that sustain each being also connect them to one another, illustrating the interconnectedness of all existence.

Purpose of Existence

Next, the system of the world is observed, extending from the level of elementary particles to the vast expanse of the universe, with a focus on their purpose.

Elementary particles exist for the purpose of forming atoms, but at the same time, they maintain their own existence as elementary particles. Atoms exist for the purpose of forming molecules, but at the same time, they maintain their own existence as atoms. Molecules exist for the purpose of forming cells and matter, but at the same time, they maintain their own existence as molecules. Cells exist for the purpose of forming tissues and organs, but at the same time, they maintain their own existence as cells. Atoms and molecules also exist for the purpose of forming minerals, which form all material bodies, such as the earth. "The earth exists for the purpose of forming the solar system, but at the same time, it maintains its own existence as the earth. The solar system exists for the purpose of forming the galaxy, but at the same time, it maintains its own existence as the solar system. The galaxy exists for the purpose of forming the universe, but at the same time, it maintains its own existence as the galaxy" (Lee, 2014, p. 119).

Therefore, it boils down to the point that all things exist with dual purposes – individual and whole. The individual purpose is to maintain their existence and development, while the whole purpose is to be part of a higher-level being and to live for the existence and development of the whole. Therefore, it can be said that all created beings, from elementary particles to the universe and to human beings, exist as connected beings with dual purposes.

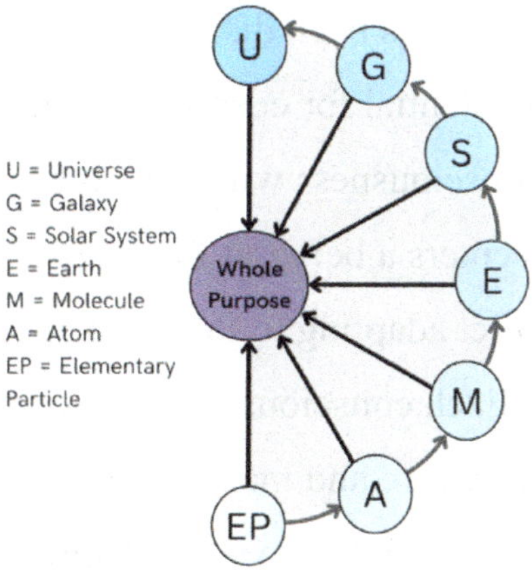

Figure 4.4 The System of Purposes for the Whole in Created Beings.

Adapted from *New Essentials of Unification Thought* by S. Lee, 2014, Unification Thought Institute (p. 120). Copyright 2014 by Unification Thought Institute.

Through the work of cosmic consciousness, all beings are believed to possess consciousness that is aware of a dual purpose. These beings are interconnected, bound by this dual purpose, and exist and develop for the collective good. This interconnectedness may align with what Dr. Bohm identified as Holomovement. It can be concluded that all things came into existence from an intended motivation and purpose that was established from the very beginning. The hidden order of the universe suggests that all beings exist to serve a higher purpose or contribute to the existence of higher beings.

2.3 Fundamental Cause of Environmental Issues

From what has been discussed so far, it points to how there is a potential for cosmic consciousness to be the origin of the consciousness within each being. Once cosmic consciousness enters a being, it functions as consciousness or internal nature, adapting to the level of complexity in its external form. Each consciousness carries a clear purpose and direction to create and maintain the interconnectedness among all beings, which is expressed through the ecological cycles that govern our world.

Based on that understanding, the fundamental cause of environmental issues led by human behavior will be addressed.

Environmental issues can be understood as disruptions or damage to ecological cycles driven by human behaviors. These behaviors are a byproduct of human consciousness, which, as discussed earlier in this chapter, is shaped by the structures of the mind and its behavior. Consequently, the root cause of these environmental issues lies in the current state of human consciousness. The human consciousness is disconnected from the cosmic consciousness, and that is what is preventing alignment with the shared purpose and direction that all other beings follow. This disconnection interferes with the natural flow of the universe, disturbing the delicate balance that sustains life on Earth.

In other words, the disconnection from the cosmic consciousness has caused the emotional impulse to seek joy in the core of the heart to become distorted into egotism or selfishness, leading to a self-centered dominance in relationships. In addition to this, this disconnection caused a lack of concern for the environment and the well-being of others can be seen among people. This disregard has contributed to numerous problems, such as environmental destruction and pollution.

As a result, efforts to address environmental issues have not been exactly smooth sailing. As mentioned in Chapter 2, many individuals prioritize their personal interests, and the same can be said regarding nations as well. Therefore, the realignment of the human consciousness with the cosmic consciousness becomes a key factor in fully addressing the environmental challenges. It is only through this alignment that harmony can be restored, not only within human beings but also with the broader ecological systems that sustain life.

Through this exploration, it has been observed that all beings exist to serve a higher being or a greater purpose. The question now arises as to whether the current state of human consciousness can align with this order. If human beings were to live for a higher purpose – for others, for nations, and for the world – while overcoming their own self-interests, what outcomes might be expected? Such a shift could lead to a world without wars and to corporations refraining from polluting water and air through an awareness of the risks to human and environmental health.

Take a moment to consider how much environmental damage is caused by war. Whenever there is a large-scale human conflict, issues such as deforestation, loss of wildlife, soil degradation, and water pollution have been associated with it. During World War I, 1.45 billion artillery shells were fired toward forests, which led to a decrease in the diversity of tree species. During World War II, 7.5 million songbirds were hunted annually in Japan for food and feathers. It is said that heavy metal contamination on the front lines of World War I is still detectable a century after the war. 5,000 tons of chemical warfare agents were dumped in the German North and Baltic Seas during World War II as well.

Throughout human history, numerous wars and conflicts have occurred despite countless efforts to prevent them and bring peace. However, conflict and war persist. Individual action to slow the environmental damage has shown to be challenging as well, as they are often limited by a lack of personal convenience and funds.

With the current state of human consciousness, this is the limitation of human effort alone. As a result of humanity's disconnection from cosmic consciousness, and with their immense capability to impact all things, it has led to extensive damage to the environment. Human beings are the only ones interfering with the interconnectedness and the direction of the universe. Therefore, it is proposed once again that in order to change these harmful behaviors and prevent further environmental damage, it is essential to identify the root cause of such behaviors within the structure of the mind and to redefine human consciousness in alignment with the cosmic consciousness.

2.4 Humanity's Unique Position

The Position of Human Beings that Enabled the Disruption of the Order and Interconnectedness of the Universe

First, the complexity of the human body and mind will be explored in order to understand humanity's position in the universe. In the diagram, "Layer Structure of Internal Nature and External Form in Existing Beings," it can be seen how humans are at the highest complexity level. This is particularly evident in the structure of the human mind, which integrates three faculties – emotion, intellect, and will. Not only the mind, but the complexity of the human body is equally remarkable. Humans have a larger brain relative to their body size compared to most animals. This allows them to process and activate a higher level of thinking. These faculties of mind and brain allow humans to possess unique abilities unseen in other beings.

For instance, humans rely on complex languages for communication. They've developed hundreds of different types of languages along with differences in culture and tradition, which highlights their intricate social structure and complexity in communication. In contrast, animals mainly rely on body language and limited vocalizations, as well as following a more simple social behavior. Consider the usage of tools. Humans have the ability to build and utilize these items to build even more things. This is a trait not commonly seen in other animals. In addition to this, humans have the ability for self-introspection as well as being more self-aware. All of these points to how humans hold a deeper cognitive complexity

Furthermore, the human body is composed of different biological systems working together in order to maintain life and respond to environmental changes. Systems such as the neurological, circulatory, respiratory, etc, all work together to make sure the human body functions properly. Bipedalism is yet another unique characteristic that humans have. Their highly developed hands allow them to perform intricate tasks, a characteristic less prominent in animals. Both in the complexity of their mind and body, it can be said that human beings stand at the highest level of all things in the universe.

The human body can also be viewed as a complex organism made from elements and life forms ranging from particles to plants and animals. These elements play an essential role for the body to function, and to live comfortably. Once again, this highlights humanity's deep connection with the natural world as well. Based on these understandings, human cognition and creativity will be explored in depth to examine human's position relative to all other entities.

Human Cognitive Abilities and Creativity in the Natural World

Humans were gifted with the cognitive ability to find and identify traits from different entities. This skill can be largely attributed to the blend of perception, memory, and analysis working together so that humans are able to discern the subtle differences in all other beings. Research in evolutionary psychology and cognitive neuroscience supports this by saying that humans have a natural ability in distinguishing between species and individuals (Mourtzi & Kanwisher, 2000). Specialized brain regions help us recognize faces and features, making this ability vital in fields like taxonomy, ethology, and anthropology. Studies in comparative cognition suggest similar findings. Humans are able to recognize and remember faces, even if they are seemingly similar individuals within a species such as primates (Haan & Nelson, 2002).

On top of that, humans can distinguish individual uniqueness within the same entities by recognizing individual traits within the same species through variations in vocal patterns, behaviors, or slight morphologies. It can be said that this skill plays a key role in social interactions; it creates opportunities for relationships as well as the formation of social hierarchies. With this skill, humans have successfully applied it to different fields, such as cultural establishments and societal systems, which ultimately led to advancements in the fields of science and medicine.

Another core ability that humans have is creativity. A skill central to innovation and technology, humans have the ability to utilize various types of materials or entities, and use it to craft many different things. Mastery over materials also means they have learned to cultivate their complex sensory perception, analytical reasoning, as well as creative problem-solving. By assessing the different properties that certain types of materials have, humans have been able to select and manipulate them for specific purposes. Research in material science supports this best when it indicates how the utilization and analysis of materials have led humans from using the simplest of tools into advanced technology that can be seen today (Ashby, 2009). These findings are supported by archaeological studies as well, when it revealed that early humans already had a sophisticated understanding of materials, using stones for tools (Dunbar, 1988; Braun et al., 2009).

Modern material science continues to build on this knowledge, allowing researchers to manipulate material properties at the molecular level. This creativity for material mastery has driven technological and cultural advancements in fields such as science, engineering, and art as human creativity.

Give and Receive Actions Between Human Beings and the Natural World

The cognitive and creative abilities that the complexity of the human mind and body houses, has been discussed. These abilities create a continuous, dynamic relationship with the natural world. This relationship can be described as *give and receive action.*

Once again, give and receive action is a dynamic process in which a correlative relationship is formed between a subject and an object, centered around a common purpose. This can either happen within a single entity, or between different beings. Through this interaction, there comes about an action of giving and receiving a certain element or force.

Figure 4.5 *Give and Receive Action*

Note. This diagram serves as a visual representation of the give and receive action. Created by the author.

Here, human beings who have an interest for other beings and seek value through education, taste, and creativity, stand in the position of the subject. The natural world, whether as a whole or represented by individual beings within it, stands in the position of the object. They are equipped with purpose and possess content (attributes). In addition, the entities standing in the object position (animals, plants, and minerals) respond to the human mind.

For instance, in the case of plants, the Backster Effect was observed by Clive Backster, an American lie-detector technician. Furthermore, Dr. Bohm stated that there may exist a certain perceptive ability even in the realms of minerals and elementary particles.

"There may be a sort of living energy in all matter that manifests in us in certain ways which it does not do in the rock. If that were the case, if a sort of intelligence were generalized throughout nature, then the speculative proposal that inanimate matter might respond to our thought is not so illogical."

(Bohm, as cited in Wilber, p. 211)

With the deep connection of human beings to the natural world, nature also continuously stimulates the human faculties of emotion, intellect, and will. This then shapes how humans perceive and respond to the world, inspiring new ideas and driving new purposeful actions.

Emotional Impact: Exposure to nature has shown to improve mental health by reducing stress and promoting healing (Kaplan & Kaplan, 1989). The Divine Principle expresses this well when it states, "We become intoxicated with the beauty of the natural world and experience the rapture of mystical union" (Moon & Eu, 2006, p. 30). For instance, the sight of a forest or the voice of the calm ocean waves can evoke feelings of awe and tranquility. This can create a sense of emotional connection with the environment. In addition to that, nature's cycle, such as the changing seasons, provides a sense of continuity and emotional stability.

Intellectual Impact: Nature can be a stimuli for intellectual curiosity and inspire creativity. The field of biomimicry demonstrated how nature can serve as the inspirational foundation for technological innovations. For example, some energy-efficient systems were built based on the structure of a tree (Cohen et al., 2012), as well as the strength of spider silk (Pezzotti et al., 2010). The complexity of natural systems encourages intellectual exploration. It helped pave the way for advancements in different scientific fields, such as physics, biology, and medicine (Vogel, 2003). The challenge of understanding nature – from the smallest particles to the vast galaxies – drives human knowledge and scientific progress (Koch & Bansal, 2014).

Willful Impact: The natural world also prompts the human will to act. It may seem like actions to protect the environment is a volitional act on the human's end. While to some extent it may be true, it may also be said that the natural world may be making an impact on humans to prompt them to take action. Issues such as climate change, biodiversity loss, and food insecurity have spurred humans to take action (McMichael et al., 2006). These challenges have induced humans to have a deep sense of responsibility and urge them to utilize adaptive strategies to address them (Rockstrom et al., 2009).

Humans have Lost the Common Purpose

The key point with this give and receive action is that the subject and object must share a common purpose. It is only when the give and receive action is centered on a shared purpose that it can generate continuous power and foster harmony. In oriental medicine, for example, there is a common purpose of "healing the human body" that creates harmony between human beings and oriental medicines. Research indicates that through specific practices in oriental medicine, such as acupuncture, the human body's power to heal is enhanced. Their effects are manifested as alleviation of gastrointestinal problems and internally the reduction of depression and anxiety (Lee et al., 2022).

As previously mentioned, all beings in the natural world, from the smallest particles to the entire universe, share the same whole purpose as the work of cosmic consciousness. However, humanity has lost this purpose due to its disconnection from cosmic consciousness, which has resulted in human behavior that interferes with ecosystems. This suggests that there is no common purpose between humanity and the natural world, even though humans still stand in the subject's position. This has made it possible for humanity to cause significant damage to the natural world.

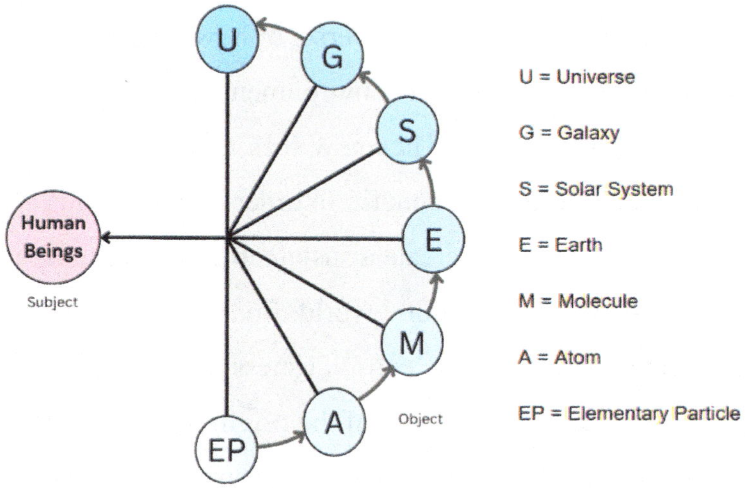

Figure 4.6 *The System of Purposes for the Whole in Created Beings.*

Adapted from New Essentials of Unification Thought by S. Lee, 2014, Unification Thought Institute (p. 120). Copyright 2014 by Unification Thought Institute.

3. Discovering the First Cause of Consciousness

From the discussion above, it has been established that due to the wrong structure of the mind, it has led to the wrong human behavior, and ultimately damaging the environment. The structure of human's mind is distorted due to the disconnection from the cosmic consciousness. Their consciousness has become unaware of the interconnectedness of the universe, or how their behaviors can impact the surrounding environment. The mind becomes self-centered; They now seek their own immediate interest and benefit. In order to change this structure of mind and create a sustainable and harmonious relationship with the natural world, finding the original connection to the cosmic consciousness is urged. Moreover, an understanding of the origin of the cosmic consciousness and its relationship to human beings.

To facilitate this change in mindset, reference may be made to Dr. Bohm's theory of the Explicate and Implicate Order, as presented by Dr. William Keepin during a lecture series at PWPA-LA (Keepin, 2024).

Explicate and Implicate Orders Proposed by Dr. Bohm

Simply defined, explicate order is directly perceptible to the mind and senses. Implicate order is all the rest of reality. The implicate reality holds these qualities:

- Present everywhere, but visible nowhere
- Extends throughout space and time, but also far beyond space and time
- Represents the unseen, that which is neither manifest nor accessible to the mind and five senses
- Has multiple levels (super-implicate orders)

Dr. Keepin provides an example to illustrate this.

There is a simple mathematical equation, $Z = Z^2 + C$, that creates explicit order in the form of a fractal structure. He suggests that there is an implicate order in the mathematical infinite. Similarly, he proposed that there is a spiritual infinite and Holomovement in the creation of fractals, which is the manifestation of the spiritual infinite through the Spiritual Seed. Dr. Keepin also suggested that consciousness itself has a fractal structure.

3.1 A Fundamental Shift in Approach

Paradigm Shift of Science with Reality and Unlock the Healing Power in Nature

Although most scientists have focused on physical reality, some have also discussed the concept of nonlocal reality, such as Dr. David Bohm. Another prime example is the research of dark matter by Dr. Zwicky. He explained how the movement of the galaxies cannot be explained with the visible mass and their gravitational pull, leading him to the hypothesis that there is black matter. However, since this was not observable with their technology, it was not widely accepted (NASA, 2023). It was only in the 1970s, by Dr. Vera Rubin, when the existence of dark matter was successfully confirmed. She measured the galaxies' rotation and compared the observed and predicted velocities of them. Her research indicated a significant discrepancy between the two, suggesting that the galaxies are embedded in a vast halo of invisible dark matter (American Museum of Natural History, 2020).

Additionally, in the late 1990s, certain observations were made. Astronomers noticed that the expansion of the universe was accelerating. It was suggested that an unknown energy, later defined as dark energy, is what drove this expansion. This led to the current understanding of dark energy. This term was coined by cosmologist Michael Turner, making a significant development in the understanding of the universe.

With these recent discoveries, along with the discovery of quantum entanglement, more and more scientists are now turning their attention to exploring the reality beyond the realm of classical physics.

Paradigm Shift of Science and Technology

Isaac Newton found gravity, and shortly after that the steam engine was invented and changed people's lives. Newton's work led to the creation of the scientific method, based on experimentation and mathematical reasoning, which became the foundation for modern science and technology. His laws of motion and gravitation (published in 1687) ushered in paradigm shifts in fields such as engineering, astronomy, and physics. These shifts allowed subsequent generations to innovate in ways that were previously unimaginable, from the steam engine in the 18th century to the modern computer in the 20th century. Together with a few other major scientific discoveries from the late 16th to the 20th centuries, this transformation is called the *Scientific Revolution*.

It is not an understatement to say that the development of the steam and gasoline engines largely were responsible for making the Industrial Revolution possible. The development of this technology marked a major paradigm shift that revolutionized what was thought to be possible. Firstly, the creation of the steam engine completely transformed how machines could be used for industrial production as well as transportation. The steam engine is a heat engine that uses steam as its primary working fluid in order to generate power that can push the pistons back and forth. The development of this technology can be attributed to an understanding of Newton's laws of motion. The application of these laws created the framework for engineers to develop technology that can convert motion into mechanical power. This engine directly contributed to the rise of factories, as well as transforming the global economy. The creation of the gasoline engine further accelerated the strengthening and presence of industrial technology as well. In 1885, Karl Benz invented the first automobile and opened a world of possibility in terms of transportation and economy. This development once again goes back to the application and understanding of Newtonian physics, which inspired the development of internal combustion engines.

Einstein then brought another level of understanding of the universe with the theory of relativity. These were the paradigm shifts that completely changed the perception of reality. Einstein's theories further pushed the boundaries of physics, breaking away from the established Newtonian view of the universe. His Special Theory of Relativity (published in 1905) and General Theory of Relativity (published in 1915) led to a paradigm shift in the understanding of space, time, and energy. His work not only changed the course of theoretical physics but also paved the way for innovations that have had practical applications in technology and space exploration, especially in systems like GPS, which rely on his theories. In the 19th and early 20th centuries, the development of many foundational theories in physics, chemistry, biology, and other sciences could be witnessed, which drastically transformed the understanding of the natural world. This was another revolutionary scientific era following the earlier Scientific Revolution. Some people even consider this to be the second scientific revolution.

Despite its successes, Einstein famously expressed his belief that quantum mechanics was not a complete theory of nature: "Quantum theory is certainly imposing. But an inner voice tells me that it is not yet the real thing.

Quantum theory says a lot, but does not bring us any closer to the secret of the Old One. I, at any rate, am convinced that He (God) does not throw dice" (Diem-Lane, 2016, p. 17).

Now is the time for a New Paradigm Shift of Understanding Reality

With the discovery of quantum entanglement, and as advancements are made in understanding dark matter, dark energy, and the quantum world, the current era is a time where a new paradigm shift of the understanding of reality is taking place – one that embraces not only the physical but the conscious aspects of existence. This shift also suggests that reality is not a static entity, but rather a dynamic, interconnected flow, aligned with the work of cosmic consciousness, striving toward the whole purpose of existence.

Currently, there is an urgent demand for sustainable energy solutions and new engine systems. With the development of engines, human processes have often come at a tremendous cost to the environment. However, with a deeper understanding of the first cause of all reality, and with humanity aligning their hearts with cosmic consciousness, it is possible that this new understanding of reality can help discover ways to harness energy that not only meets humanity's needs but also preserves the health of the Earth.

The boundaries of science are being re-drawn beyond the confines of time and space to explore the unseen realms of reality. By connecting the core of a human's heart and mind to connect to the first cause, it will allow them to awaken their collective consciousness. This new paradigm shift will guide humanity towards a future where the gap between science and religion can be bridged by changing the structure of the mind and behavior. This transformation will enable humans to discover new ways to make energy sustainable and create a pleasant living environment. Ultimately, it will lead to creating harmony between technology and Earth.

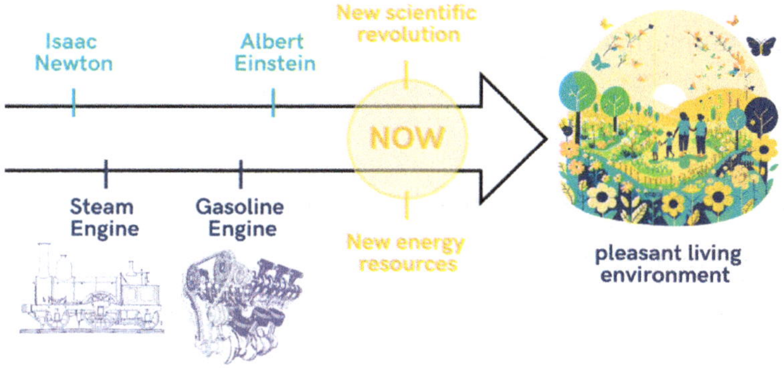

***Figure 4.7** Paradigm Shift towards a pleasant living environment*

Note. This diagram visually represents major paradigm shifts, emphasizing how a new one can contribute to a pleasant living environment. Created by the author.

4. Stewardship of Love to Unlock the Potential in Nature

4.1 Nature's Inherent Ability to Self Restore

Nature inherently possesses the capacity for self-restoration through various natural processes. However, if humans shift the structure of their minds by reconnecting their consciousness with the cosmic consciousness—aligning with its latent purpose—this healing power could be greatly amplified. Such a shift would enhance nature's restorative abilities, allowing for a more profound and effective healing process than what is currently happening.

Below are some key examples from scientific studies that demonstrate how natural systems already possess restorative powers and play crucial roles in environmental recovery.

Wetlands as Natural Water Filtration Systems

Wetlands are considered to be one of the most productive ecosystems in the world. Commonly deemed as the kidneys of the Earth, they serve the function of filtering and purifying the water by removing pollutants and sediments before it reaches other bodies of water. As water travels from a higher elevation to a lower area of water, it picks up a lot of pollutants as well as chemicals along the way. Through the process of sediment trapping, nutrient removal by plants and microorganisms, and chemical detoxification as well as slowing the flow of water, wetlands serve as a naturally efficient filter to clean out the water.

On top of that, wetlands also serve as a habitat for a wide variety of species, making it an extremely rich biodiverse ecosystem. Wetlands store carbon within their plant communities as well, contributing to carbon sequestration. This process helps control climate change and the overall CO2 presence in the atmosphere. During times of a storm or other natural disaster as well, wetlands serve as a flood buffer by slowing down the flow of water. This process of receiving water, absorbing the excess, and slowly releasing it significantly reduces the risk of downstream flooding. Wetlands play a major role in promoting environmental health, and they may serve as inspiration for how nature restores and maintains itself (US EPA, 2019).

Phytoremediation and Natural Detoxification

Phytoremediation is defined as the use of green plants to remove pollutants from the environment or render them harmless. Primarily, toxic heavy metals and organic pollutants are seen as the targets for phytoremediation. Plants such as sunflowers, Indian mustards, and even corn have the capacity to store high concentrations of lead. These plants are naturally able to absorb and remove certain pollutants to cleanse the environment (Salt et al., 1998).

Phytoremediation appears to be a potential solution towards bolstering the efforts of detoxifying the Earth that has been contaminated by pollutants and human waste. Studies have shown that wastes such as heavy metals, organic contaminants, antibiotics, and pesticides can all be remediated by different types of plants. The utilization of these green plant's ability to accumulate elements and compounds from the environment and to perform biochemical transformations may serve as a new frontier of plant biology (Salt et al., 1998).

Marine Ecosystem Restoration

Marine ecosystems, most notably the coral reefs, are known to have remarkable restorative and regenerative ability. Coral Reefs undergo a process called Coral Spawning, where female and male corals release reproductive cells into the water as a synchronized event.

This allows the ecosystem to recover and maintain its functions over time.

In addition to natural recovery, the establishment of marine protected areas (MPAs) enhances coral reef restoration efforts. These areas help reduce human impacts by limiting activities such as fishing and pollution, which aids in the natural recovery of marine habitats. Marine ecosystems such as coral reefs continue to play an important role in climate regulation as well as provide a habitat for many marine species. The implementation of these marine protected areas are an essential step towards promoting the recovery of these ecosystems (McLoed et al., 2009).

Soil Regeneration and Microbial Communities

Soil regeneration is a form of ecological regeneration where the focus lies on creating new and rejuvenating soil health. In this recovery process, the microbial communities play a central role in facilitating the decomposition of organic matter, suppressing the growth of pathogens, as well as releasing nutrients into the soil. This then also translates to the establishment of plant species as well.

With regards to this, Dr. Lal discusses the importance of microbial activities in the soil which directly translates to soil resilience. Soil resilience is the ability of the soil to recover its quality in response to changes (Lal, 2015). The identification of systems that stimulate soil microbiotic activity and related microbial processes can translate to reducing the risk for soil degradation. Additionally, it is stated that regenerative agriculture and agro-ecology are important concepts to consider which aim to restore soil. Techniques such as agroforestry, mixed farming systems, and implementation of strategies that integrate soil fertility management will help to restore and let the soil thrive (Lal, 2023).

4.2 Technological Solutions and Challenges

Extensive human effort has already been made to create technologies that enhance nature's innate restorative abilities to reverse some of the environmental damages. These efforts may still be challenged by issues such as effectiveness, efficiency under different environmental conditions, adaptability to specific types of pollutants, or other unknown indirect environmental consequences. There is also the concern of over reliance on technological solutions diverting the attention from the essential systemic changes needed in human behavior to resolve the root causes of environmental degradation. However, the human effort to tackle these environmental issues must be acknowledged.

Drone and Remote Sensing Technology

Drone and satellite imaging technologies are starting to play a much bigger role in monitoring ecosystems as well as keeping track of restorative efforts. Due to their ability to cover a wide area as well as remotely collect data, they are able to provide real-time data on soil conditions, vegetation growth and even changes in biodiversity. In addition to this, drones are also being used for places and situations that may be limiting for a physical body. In areas that are geographically inaccessible for people, aerial seeding by drones are an intervention for ecological support (Anderson & Gaston, 2013).

Carbon Capture and Storage (CCS) for Carbon Sequestration

CCS is a method of reducing CO2 emissions which plays an important role in addressing human-induced climate change. Briefly explained, this is a process where CO2 emissions from industrial activity are captured and stored underground. It essentially transports and moves the

CO_2 from the atmosphere. On a large scale, this method of CO_2 sequestration may aid with reducing the effects of climate change (IPCC, 2005).

Artificial Photosynthesis

Artificial photosynthesis is a process by which sunlight is converted into storable energy forms such as hydrogen or other solar fuels. This is done by developing cells that can split water into their atomic components or by utilizing solar energy to reduce the amount of CO_2 in the atmosphere. Through this, it holds the potential to produce energy-dense fuels, which may lead to a sustainable solution for energy consumption and energy security. Through this process, it may reduce atmospheric CO_2 levels while also producing clean energy (Machín et al., 2023).

4.3 Nature-Based Solutions (NbS)

Together with these engineered solution technologies, great potential can be seen in Nature-Based Solutions (hereafter abbreviated as NbS). NbS are strategies that focus on working with nature; they utilize natural processes and ecosystems in order to resolve environmental challenges as well as provide a beneficial return for humans. They aim to restore or protect ecosystems to tackle issues such as water management, disaster resilience, and biodiversity loss. NbS includes a variety of strategies like reforestation, wetland restoration, sustainable agriculture, and the creation of urban green spaces. By harnessing the power of nature, these approaches are able to address and manage some of the environmental issues mentioned. Listed below are some examples of these approaches.

Agroforestry

Agroforestry is the intentional integration of trees and shrubs into agricultural systems to create environmental, economic, and social benefits. For example, planting trees on farms not only provides shade and enhances soil fertility, but it also helps reduce soil erosion, sequester carbon, and improve water retention in agricultural land (Lal, 2015).

Revegetation of Degraded Land

The revegetation of degraded land focusing on invigorating vegetative efforts to areas that have been impacted by human activities to rebuild the soil of disturbed land. Practices such as the expansion of grassland and reforestation can increase biodiversity, absorb atmospheric carbon dioxide and control soil erosion (Zhang et al., 2024).

Living Shorelines

Living shorelines use natural materials such as plants, rocks, and sand to stabilize shorelines and prevent erosion as an alternative to relying on hard structures like seawalls. For instance, planting salt-tolerant vegetation along coastal areas or installing oyster reefs can reduce wave energy and help prevent coastal erosion (Palinkas et al., 2023).

NbS has been recognized at the global level with the UN initiatives aiming to promote and scale the adoption of these solutions. In the recent decade, a number of initiatives can be seen from the UN in an effort to develop solutions that align with the Sustainable Development Goals (SDG), primarily focusing on climate change, life on land, and life below water. In 2011, the UN Environment Programme's ecosystems-based adaptation approach was initiated in order to help communities adapt to climate change by utilizing NbS and ecosystem services. In addition to that, the UN Decade on Ecosystem Restoration was launched in 2021 in order to prevent the degradation of ecosystems and create a sustainable future. With this, NbS is considered a crucial tool in mitigating climate change.

However, similar to those faced by engineered solutions, the application of these approaches may be faced with several obstacles. Most of the NbSs, such as reforestation and wetland restoration, take time to yield results. Solutions that require a substantial amount of time may be subjected to short-term political or economic pressures, making them difficult to maintain and sustain. In addition to that, these solutions have a lot of moving pieces, which require close coordination of different stakeholders. In areas where resources are scarce, the initial cost of investment and financial restraints in regard to application of these solutions may also serve as a barrier as well. With these points in mind, collaborative efforts among the government as well as the local community become essential. In addition to that, a long-term commitment becomes a key part in making this a success. However, the starting point for all these ideas to be put into motion is increased awareness. There must be the willingness of the people to embrace these approaches in order to make them a reality.

4.4 Amplifying Power Through Interconnected Stewardship of Love

These examples illustrate the remarkable restorative abilities of nature and the cognitive and creative abilities of humans, which led to the development of engineered solution technologies and NbS. The combined potential of nature's inherent restorative powers and human innovation demonstrates the ability to heal the environment more effectively and faster than ever expected.

When these points are considered, it becomes clear that the core challenges are not about knowledge or technology deficits. The true challenge is fostering a collective will and commitment among people to make it happen. It is the willingness to embrace these solutions, long-term dedication, and compassion to work together for the well-being of the planet that will ultimately drive meaningful change. This requires a shift in the hearts of people, as well as a shift in the structure of the mind.

While natural processes are already powerful, they can be further strengthened when humans align their consciousness with the greater cosmic purpose. By changing the structure of the mind, human consciousness will recognize the interconnectedness of all life and support ecological integrity. Humanity's relationship with nature must change. It must shit from one based on self-centered domination to stewardship rooted in love.

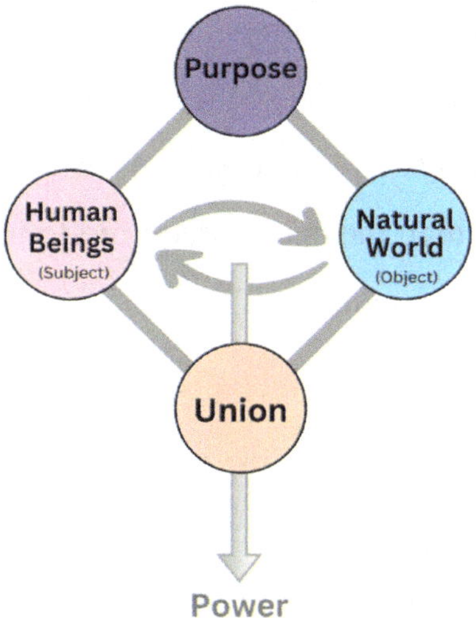

Figure 4.8 *Give and Receive Action and the Four Position Foundation.*

Adapted from *New Essentials of Unification Thought* by S. Lee, 2014, Unification Thought Institute (p. 44). Copyright 2014 by Unification Thought Institute.

This give and receive action between human beings and the natural world will ignite a dynamic force, accelerating at an extraordinary pace. As a result, this newfound power will speed the healing process of the natural world. The nature of give and receive action is that it will produce a stronger output of power than the initial input through the exchange of power/elements between the subject and the object. This can already be observed in nature.

- **Photosynthesis (Biological Systems)**. Through the exchange of sunlight, water, and CO2, plants produce glucose and oxygen, generating more energy than the initial input (Taiz et al., 2015)
- **Nuclear Fusion (physics)**. Through the exchange of hydrogen nuclei, which fuses to produce helium, it produces a tremendous amount of energy, far greater than the energy required to initiate the process (Griem, 2012)
- **Hydroelectric Power Generation (Mechanical Systems)**. Through the exchange of gravitational potential energy of water, it produces electricity with the output being greater than the energy required to move the water (Mujumdar, 2013)
- **Magnetic Induction (Electromagnetic Systems)**. Through the exchange of mechanical energy with the movement of magnets in a coil of wire, it generates electrical energy, creating more power than the mechanical input (Fitzpatrick, 2013)
- **Convection Currents (Atmospheric and Oceanic Systems)**. Through the exchange of heat energy, warm air or water rises while cooler air or water sinks, generating currents that can be harnessed for

power, with the output exceeding the input heat energy from the sun (Malkus & Veronis, 1958)

- **Geothermal Energy (Thermal Systems)**. Through the exchange of heat from the Earth's interior, it produces electrical energy, generating more power than the thermal input (Lund & Boyd, 2016)

Therefore, by applying the same principle between humanity and the natural world, nature's capacity for healing can be amplified. If humanity can connect its consciousness to the Cosmic Consciousness, understanding that a shared purpose links all life, it may uncover untapped potential within both humanity and the natural world. This connection could allow us to restore the environment more rapidly than previously imagined.

We look forward to scientific discoveries that could transform our understanding of reality—especially the possibility of a nonlocal being, a First Cause that possesses consciousness and an inherent desire to seek joy, much like humanity itself. This First Cause is believed to hold the origin of all forms of energy, as well as the mathematical laws that govern the universe and everyday life. With this new understanding, we can transform our consciousness and approach to the world.

When the potential of humans and the natural world are combined, the result won't simply be their sum, but rather a powerful, transformative product. Through this shift in consciousness and purpose, we shall heal the damage we've caused and create a previously unseen, beautiful environment. This harmonious collaboration will be the fundamental solution to the environmental issues we have created.

Epilogue

Integrating the Ways of Religion and Science in a Unified Endeavor

To conclude this innovative solution for environmental issues, we will delve deeper into the importance of the unity of religion and science, which lies at the heart of our founders' vision and is seen as the only fundamental solution to all the social issues we have faced historically.

To find the fundamental solution, we have focused on the interconnectedness of all things in the universe as the work of consciousness. This interconnectedness has been recognized by science, such as ecology and systems theory. These highlight the intricate connections within ecosystems, the web of life, and the cosmos. At the same time, a similar idea has been recognized by religion, which often speaks of a divine interconnectedness where everything is part of a greater cosmic plan.

This similarity requires that we take a step back and consider their shared purpose, which is to seek truth and understanding. We find that these two domains of knowledge should be integrated into one unified undertaking, even though religion and science have always been placed on opposite ends of the spectrum in the modern era.

The reason for the separation between religion and science may be found in modern scientific inquiry, grounded in empiricism. Empiricism asserts that knowledge is obtained only through sensory experience (sight, hearing, touch, etc.). This empirical approach to science stemmed from the Scientific Revolution and the Enlightenment. The revolutionary scientific discoveries made by key scientists like Galileo Galilei, Isaac Newton, and Johannes Kepler influenced Enlightenment thinkers. Francis Bacon laid the foundation for empiricism, John Locke systematized it, and David Hume drew logical conclusions from it. Hume even brought empiricism to the realm of skepticism and doubted the very existence of the spiritual substance, considering it to be nothing more than a bundle of perceptions (Lee, 2014).

However, the two fields may not be as far apart as we think they are. Many of these scientists saw their work as a way to understand not just the mechanics of nature but also the broader expression of higher reason. Galileo Galilei, Isaac Newton, and Johannes Kepler all left behind words to express their views on science and faith.

"I do not feel obliged to believe that the same God who has endowed us with sense, reason, and intellect has intended us to forego their use..."

—Galileo Galilei (Galilei, 1615).

"This most beautiful system of the sun, planets, and comets, could only proceed from the counsel and dominion of an intelligent and powerful Being. And that Being governs all things, not as the soul of the world, but as Lord over all; and on account of his dominion, he is wont to be called Lord God, Universal Ruler."

—Isaac Newton (Jourdain, 1915, p. 247).

"Geometry, coeternal with God and shining in the divine Mind, gave God the pattern... by which he laid out the world so that it might be best and most beautiful and finally most like the Creator."

—Johannes Kepler (Kepler, 1619, as quoted in Field, 1988, p. 123).

Those words suggested that they combined their belief in God with their scientific endeavors, seeing the two as complementary rather than conflicting.

Religion has tried to discern what is good and evil, the nature of knowledge, and to address questions about ethics, which are important considerations in scientific practices. With the recent rapid advancement of certain areas, such as biotechnology, medicine, environmental science, and communication technology, scientists must navigate the moral considerations of their discoveries more than ever. In this modern age, many scientists have limited themselves to the pursuit of objective facts, disregarding values of any kind. The result is the chaotic situation we see today. The ethical implications of genetic engineering, nuclear technology, and artificial intelligence are more necessary than ever. Our founders, Rev. Sun Myung Moon and Dr. Hak Ja Han Moon, expressed their hope for scientists to first be persons of values, or persons of character, before being scientists, so that ethics can become the basis of natural science. They sponsored the International Conferences on the Unity of the Sciences (ICUS), together with the Professors World Peace Academy conferences, to encourage scientists and scholars to address values, so that they might restore true creativity. We manifest 'true creativity under the ethic which requires us to love nature, to reconsider the dignity of human beings, to seek love among all humankind, and to search for God as the origin of love' (Lee, 2014).

The true potential of human understanding lies in the integration of science and religion, where both knowledge come together in full consonance. Rather than viewing them as separate endeavors, we should see them as complementary pairs, like the human mind and body. With this integration, we can awaken our consciousness to approach the natural world with a sense of love and respect. We will understand that our actions are part of a larger, interconnected whole. This fundamental change in human consciousness can guide us to live more sustainably and ethically. We shall foster harmony between humanity and the natural world.

Science provides the tools to understand the natural world, while religion provides the wisdom of the giving purpose behind it. When we embrace both, we can develop a unified worldview that is rich, holistic, and profound. This integrated approach to knowledge allows for a more comprehensive understanding of reality that honors both the physical and the spiritual, the empirical and the metaphysical, the material and the spiritual. Only then, completely liberated from ignorance and living solely in goodness in accordance with the desire of the original mind, will we enjoy eternal happiness. We will maintain a harmonious relationship with the natural world and exercise true creativity in science and technology to provide a pleasant lifestyle for all humanity.

References

New Approach

Lal, R. (2020). Managing soils for resolving the conflict between agriculture and nature: The hard talk. *European Journal of Soil Science, 71*, 1–9. https://doi.org/10.1111/ejss.12857

Environmental Impacts of Energy Production

Aka, P. C. (2024). *When the Nose Cries, the Eye Also Cries: Why Environmental Pollution in Nigeria's Niger Delta Region is a Global Problem*. PWPA Los Angeles. https://www.pwpala.org/pwpa-vi-when-the-nose-cries-the-eye-also-cries

Al Mubarak, F., Rezaee, R., & Wood, D. A. (2024). Economic, societal, and environmental impacts of available energy sources: A Review. MDPI. https://doi.org/10.3390/eng5030067

Allen, F. (2012). The Enemy Within: Oil in the Niger Delta. *World Policy Journal*, *29*(4), 46–53. http://www.jstor.org/stable/23326712

Badichek, G. (2019). Resolving Conflicts Between Endangered Species Conservation and Renewable Energy Siting: Wiggle Room for Renewables? *Consilience*, *21*, 187–210. https://www.jstor.org/stable/26775091

Bertrand, S. (2021). Fact sheet: Climate, environmental, and health impacts of fossil fuels. EESI. https://www.eesi.org/papers/view/fact-sheet-climate-environmental-and-health-impacts-of-fossil-fuels-2021#:~:text=Air%20pollution%20from%20fossil%20fuels,forests%2C%20and%20harm%20to%20wildlife

Blekhman, D. (2023). *Sun, Wind, and Water for Hydrogen*. PWPA Los Angeles. https://www.pwpala.org/pwpaforum-sun-wind-and-water-for-hydrogen

Bruederle, A., & Hodler, R. (2019). Effect of oil spills on infant mortality in Nigeria. Proceedings of the

National Academy of Sciences of the United States of America, 116(12), 5467–5471. https://doi.org/10.1073/pnas.1818303116

Ebiede, T. M. (2011). *Conflict Drivers:* Environmental Degradation and Corruption in the Niger Delta Region. *African Conflict and Peacebuilding Review*, *1*(1), 139–151. https://doi.org/10.2979/africonfpeacrevi.1.1.139

Energy Institute. (2024). *Energy Institute Statistical Review of World Energy 2024*. Statistical review of world energy. https://www.energyinst.org/statistical-review

Environmental Protection Agency. (2025a). *Carbon Dioxide Emissions*. EPA. https://www.epa.gov/ghgemissions/carbon-dioxide-emissions

Environmental Protection Agency. (2025b). *Climate Change Indicators: Greenhouse Gases*. EPA. https://www.epa.gov/climate-indicators/greenhouse-gases

Hamilton, T. (2015). House of The Rising Sun. *Corporate Knights*, *14*(3), 34–37. http://www.jstor.org/stable/44149866

Ibaba, I. S. (2011). Corruption, human-rights violation, and the interface with violence in the Niger Delta. *Development in Practice*, *21*(2), 244–254. http://www.jstor.org/stable/23048471

International Energy Agency. (2021). *The cost of capital in clean energy transitions – analysis*. IEA. https://www.iea.org/articles/the-cost-of-capital-in-clean-energy-transitions

International Energy Agency. (2023). *World energy outlook 2023*. International Energy Agency. https://www.iea.org/reports/world-energy-outlook-2023. Licence: CC BY 4.0 (report); CC BY NC SA 4.0 (Annex A).

International Renewable Energy Agency. (2023). *World Energy Transitions Outlook 2023*. IRENA. https://www.irena.org/Digital-Report/World-Energy-Transitions-Outlook-2023

Intergovernmental Panel on Climate Change. (2023). *Ar6 synthesis report: Summary for policymakers headline statements*. The Intergovernmental Panel on Climate Change. https://www.ipcc.ch/report/ar6/syr/resources/spm-headline-statements/

Lindsey, R. (2023). Climate change: Global sea level. NOAA Climate.gov. https://www.climate.gov/news-features/understanding-climate/climate-change-global-sea-level

Miller, A. (2024). Fossil fuel air pollution responsible for 1 in 5 deaths worldwide. Harvard T.H. Chan School of Public Health. https://hsph.harvard.edu/environmental-health/news/fossil-fuel-air-pollution-responsible-for-1-in-5-deaths-worldwide/

National Aeronautics and Space Administration. (2024). *Causes - NASA science*. NASA. https://science.nasa.gov/climate-change/causes/

NASA Science. (2013). *Graphic: The relentless rise of carbon dioxide*. NASA.

https://science.nasa.gov/resource/graphic-the-relentless-rise-of-carbon-dioxide/

NOAA National Centers for Environmental Information. (2024). *Monthly Global Climate Report for Annual 2023*. National Centers for Environmental Information (NCEI). https://www.ncei.noaa.gov/access/metadata/landing-page/bin/iso?id=gov.noaa.ncdc%3AC00672

National Oceanic and Atmospheric Administration. (2024). No sign of greenhouse gases increases slowing in 2023. NOAA Research. https://research.noaa.gov/no-sign-of-greenhouse-gases-increases-slowing-in-2023

Natural Resources Defense Council. (2009). "Ocean Acidification: The Other CO2 Problem". NRDC. https://www.nrdc.org/sites/default/files/NRDC-OceanAcidFSWeb.pdf.

Pitkin, J. (2013). *Oil, Oil, Everywhere: Environmental and Human Impacts of Oil Extraction in the Niger Delta*. (Pomona Senior Theses). 88. https://scholarship.claremont.edu/pomona_theses/88

Semieniuk, G., Weber, I., Weaver, I., Wasner, E., Braun, B., Holden, P. B., Salas, P., Mercure, J.-F., & Edwards, N. R. (2024). *Distributional implications and share ownership of Record Oil and gas profits*. PERI (Political Economy Research Institute). https://peri.umass.edu/publication/distributional-implications-and-share-ownership-of-record-oil-and-gas-profits/

UN Environment Programme. (n.d.). *Ogoniland's oil history*. UNEP. https://www.unep.org/topics/disasters-and-conflicts/country-presence/nigeria/ogonilands-oil-history

United Nations. (n.d.). *Causes and effects of climate change*. United Nations. https://www.un.org/en/climatechange/science/causes-effects-climate-change

US Department of Energy. (n.d.). *Doe explains...the carbon cycle*. Energy.gov. https://www.energy.gov/science/doe-explainsthe-carbon-cycle

US Energy Information Administration. (2023). *International Energy Outlook 2023*. U.S. Energy Information Administration (EIA). https://www.eia.gov/outlooks/ieo/narrative/index.php

US Geological Survey. (2022). *USGS Carbon Cycle*. USGS. https://www.usgs.gov/media/images/usgs-carbon-cycle

How City Infrastructure Contributes to Water Pollution

Aguilera, R., Gershunov, A., & Benmarhnia, T. (2019). Atmospheric rivers impact California's coastal water quality via extreme precipitation. *Science of the Total Environment, 671*, 488-494. https://doi.org/10.1016/j.scitotenv.2019.03.318

Anderson-Mahoney, P., Kotlerman, J., Takhar, H., Gray, D., & Dahlgren, J. (2008). Self-reported health effects among community residents exposed to perfluorooctanoate. *New solutions: A Journal of Environmental and Occupational Health Policy,*

18(2), 129–143. https://doi.org/10.2190/NS.18.2.d

Dupont Media. (2020). DuPont Closes Four Clean Water Technology Acquisitions. https://www.dupont.com/news/dupont-closes-four-clean-water-technology-acquisitions.html

Edgar, G. J., Stuart-Smith, R. D., Heather, F. J., Barrett, N. S., Turak, E., Sweatman, H., Emslie, M. J., Brock, D. J., Hicks, J., French, B., Baker, S. C., Howe, S. A., Jordan, A., Knott, N. A., Mooney, P., Cooper, A. T., Oh, E. S., Soler, G. A., Mellin, C., … Bates, A. E. (2023). Continent-wide declines in shallow reef life over a decade of ocean warming. *Nature*, *615*(7954), 858–865. https://doi.org/10.1038/s41586-023-05833-y

Escher, B. I., Stapleton, H. M., & Schymanski, E. L. (2020). Tracking complex mixtures of chemicals in our changing environment. *Science,* 367(6476), 388–392. https://doi.org/10.1126/science.aay6636

Ewere, E. E., White, S., Mauleon, R., & Benkendorff, K. (2024). Soil microbial communities and degradation of pesticides in greenhouse effluent through a woodchip bioreactor. *Environmental Pollution*, *359*,

124561. https://doi.org/10.1016/j.envpol.2024.124561

Gaber, N., Bero, L., & Woodruff, T. J. (2023). The Devil they Knew: Chemical Documents Analysis of Industry Influence on PFAS Science. *Annals of global health, 89*(1), 37. https://doi.org/10.5334/aogh.4013

Given, S., Pendleton, L.H., & Boehm, A. B. (2006). Regional Public Health Cost Estimates of Contaminated Coastal Waters: A Case Study of Gastroenteritis at Southern California Beaches. *Environmental Science & Technology, 40*(16), 4851-4858. https://doi.org/10.1021/es060679s

Konwinski, K. P., & Olayinka, O. (2022). PFAS: The Impact of Forever Chemicals. *The Brief, 51*(3), 32-41.

Muir, D. C. G., Getzinger, G. J., McBride, M., & Ferguson, P. L. (2023). How Many Chemicals in Commerce Have Been Analyzed in Environmental Media? A 50 Year Bibliometric Analysis. *Environmental Science & Technology, 57*(25), 9119–9129. https://doi.org/10.1021/acs.est.2c09353

Parsons, T. (2021). The Weight of Cities: Urbanization Effects on Earth's Subsurface. *AGU Advances, 2*(1), e2020AV000277. https://doi.org/10.1029/2020AV000277

Pendergraft, M.A., Belda-Ferre, P., Petras, D., Morris, C.K., Mitts, B.A., Aron, A. T., Bryant, M., Schwartz, T., Ackermann, G., Humphrey, G., Kaandorp, E., Dorrestein, P. C., Knight, R., & Prather, K. A. (2023). Bacterial and Chemical Evidence of Coastal Water Pollution from the Tijuana River in Sea Spray Aerosol. *Environmental Science & Technology, 57*(10), 4071-4081. https://doi.org/10.1021/acs.est.2c02312

Schmidt, J.T., Wu, M. S. C., Kittner, H.E., Arey, J. S., Hammond, D. E., Earth 182A Group, & Valentine, D. L. (2024). Disentangling the History of Deep Ocean Disposal for DDT and Other Industrial Waste Off Southern California. *Environmental Science & Technology, 58*(9), 4346-4356. https://doi.org/10.1021/acs.est.3c08575

Shearer, J. J., Callahan, C. L., Calafat, A. M., Huang, W. Y., Jones, R. R., Sabbisetti, V. S., Freedman, N. D., Sampson, J. N., Silverman, D. T., Purdue, M. P., & Hofmann, J. N. (2021). Serum Concentrations of

Per- and Polyfluoroalkyl Substances and Risk of Renal Cell Carcinoma. *Journal of the National Cancer Institute, 113*(5), 580–587. https://doi.org/10.1093/jnci/djaa143

Smalling, K.L., Romanok, K. M., Bradley, P. M., Morriss, M. C., Gray, J. L., Kanagay, L. K., Gordon, S. E., Williams, B. M., Breitmeyer, S. E., Jones, D. K., DeCicco, L. A., Eagles-Smith, C. A., & Wagner, T. (2023). Per- and Polyfluoroalkyl substances (PFAS) in United States Tap Water: Comparison of underserved private-well and public-supply exposures and associated health implications. *Environmental International, 178.* https://doi.org/10.1016/j.envint.2023.108033

Stenstrom, M. K. (1998). Wastewater Treatment. *Southern California Environmental Report Card 1998 / UCLA Institute of the Environment*, 17-26.

Stenstrom, M. K. (1999). Stormwater Impact. *Southern California Environmental Report Card 1999 / UCLA Institute of the Environment*, 13-21.

United Nations. (n.d.). Populations. https://www.un.org/en/global-issues/population

United States Environmental Protection Agency (US EPA). Montrose Chemical Corp. Torrance, CA *Cleanup*

Activities.
https://cumulis.epa.gov/supercpad/SiteProfiles/index.cfm?fuseaction=second.cleanup&id=0900993#content

United States Environmental Protection Agency (US EPA). (2024). 2022 TRI National Analysis. https://www.epa.gov/system/files/documents/2024-03/complete_2022_tri_national_analysis.pdf

Yeager, P. C. (1993). Industrial Water Pollution. Beyond the Law: Crime in Complex Organizations, 18, 97-148. https://doi.org/10.1086/449223

Zutshi, B., & Prasad, S.G.R. (2008). Impact of Pollution on Fresh and Marine Water Resources - A Review. *Pollution Research 27*(3), 461-466.

Soil Sciences and Agronomy in Relation to Climate Change

Annan, K. (2016). Africa and the Global Security Architecture. A Keynote Addressed Delivered as the then Chair of the African Progress Panel Held at Bahir Dar.

Bajželj, B., & Richards, K. (2014). The Positive Feedback Loop between the Impacts of Climate Change and

Agricultural Expansion and Relocation. *Land (Basel)*, *3*(3), 898–916. https://doi.org/10.3390/land3030898

Beddington, J. R., Asaduzzaman, M., Clark, M. E., Fernández Bremauntz, A., Guillou, M. D., Howlett, D. J. B., Jahn, M. M., Lin, E., Mamo, T., Negra, C., Nobre, C. A., Scholes, R. J., Van Bo, N., & Wakhungu, J. (2012). Agriculture. What next for agriculture after Durban? *Science (American Association for the Advancement of Science)*, *335*(6066), 289–290. https://doi.org/10.1126/science.1217941

Bows-Larkin, A., Dawkins, E., Gough, C., Mander, S., Mclachlan, C., Röder, M., Thom, L., Thornley, P., & Wood, R. (2012). *What's Cooking? Adaptation & Mitigation in the UK Food System*. University of Manchester, Sustainable Consumption Institute.

Brown, L. R. (Lester R. (1995). Who will feed China? : wake-up call for a small planet (First edition.). W.W. Norton & Company.

Dlamini P, Chivenge P, Chaplot V. (2016). Overgrazing decreases soil organic carbon stocks the most under day climates and low soil pH: a meta-analysis shows. Agriculture, Ecosystems and Environment

221: 258–269.
https://doi.org/10.1016/j.agee.2016.01.026

Erda, L., Wei, X., Hui, J., Yinlong, X., Yue, L., Liping, B., & Liyong, X. (2005). Climate change impacts on crop yield and quality with CO2 fertilization in China. *Philosophical Transactions of the Royal Society of London. Series B. Biological Sciences, 360*(1463), 2149–2154. https://doi.org/10.1098/rstb.2005.1743

Ford, J. D. (2009). Vulnerability of Inuit food systems to food insecurity as a consequence of climate change: a case study from Igloolik, Nunavut. *Regional Environmental Change*, *9*(2), 83–100. https://doi.org/10.1007/s10113-008-0060-x

Gomiero T. (2016). Soil degradation, land scarcity and food security: reviewing a complex challenge. *Sustainability 8*(3), 281. https://doi.org/10.3390/su8030281

Intergovernmental Panel on Climate Change (IPCC). (2023). *Climate Change 2022 – Impacts, Adaptation and Vulnerability: Working Group II Contribution to the Sixth Assessment Report of the Intergovernmental Panel on Climate Change.*

Cambridge University Press. https://doi.org/10.1017/9781009325844

Kemper, K. J., & Lal, R. (2017). Pay dirt! human health depends on soil health. Complementary Therapies in Medicine, 32, A1-A2.

Lal R. (1976). Soil erosion on Alfisols in western Nigeria. I. Effects of slope, crop rotation and residue management. *Geoderma 16*(5), 363–375. https://doi.org/10.1016/0016-7061(76)90001-X

Lal, R. (2004). Soil Carbon Sequestration Impacts on Global Climate Change and Food Security. *Science (American Association for the Advancement of Science)*, *304*(5677), 1623–1627. https://doi.org/10.1126/science.1097396

Lal, R. (2010a). Beyond Copenhagen: mitigating climate change and achieving food security through soil carbon sequestration. Food Security, 2(2), 169–177. https://doi.org/10.1007/s12571-010-0060-9

Lal, R. (2010b). Managing Soils and Ecosystems for Mitigating Anthropogenic Carbon Emissions and Advancing Global Food Security. *Bioscience*, *60*(9), 708–721. https://doi.org/10.1525/bio.2010.60.9.8

Lal, R., Faiz, S. M. A., Islam, K. R., Rahman, A. H. M. M., Sivakumar, M., Faiz, S. M. A., Islam, K. R.,

Sivakumar, M. V. K., & Mustafizur Rahman, A. H. M. (2011). Climate Change and Food Security in South Asia (1. Aufl.). Springer Science + Business Media. https://doi.org/10.1007/978-90-481-9516-9

Lal, R. (2012). Climate Change and Soil Degradation Mitigation by Sustainable Management of Soils and Other Natural Resources. Agricultural Research (India : Online), 1(3), 199–212. https://doi.org/10.1007/s40003-012-0031-9

Lal, R. (2013a). Food security in a changing climate. *Ecohydrology & Hydrobiology*, *13*(1), 8–21. https://doi.org/10.1016/j.ecohyd.2013.03.006

Lal, R. (2013b). Soil carbon management and climate change. Carbon Management, 4(4), 439–462. https://doi.org/10.4155/cmt.13.31

Lal, R. (2019). Promoting "4 Per Thousand" and "Adapting African Agriculture" by south-south cooperation: Conservation agriculture and sustainable intensification. Soil & Tillage Research, 188, 27–34. https://doi.org/10.1016/j.still.2017.12.015

Lal, R. (2023). Farming systems to return land for nature: It's all about soil health and re-carbonization of the terrestrial biosphere. *Farming System*, *1*(1), 100002. https://doi.org/10.1016/j.farsys.2023.100002

Lal, R. (2024). Soil, Spirituality, Science, and Religion. PWPA Los Angeles. https://www.pwpala.org/soil-and-soul

Kerr, R.B., Hasegawa, T., Lasco, R., Bhatt, I., Deryng, D., Farrell, A., Gurney-Smith, H., Ju, H., Lluch-Cota, S., Meza, F., Nelson, G., Neufeldt, H., & Thornton, P. (2022). Food, fibre, and other ecosystem products in *Climate Change 2022: Impacts, Adaptation and Vulnerability.* Contribution of Working Group II to the Sixth Assessment Report of the Intergovernmental Panel on Climate Change, H.-O. Pörtner, D. C. Roberts, M. Tignor, E. S. Poloczanska, K. Mintenbeck, A. Alegría, M. Craig, S. Langsdorf, S. Löschke, V. Möller, A. Okem, B. Rama, Eds. Cambridge Univ. Press. chap. 5, pp. 713–906.

R. Lal, M.V.K. Sivakumar, S.M.A. Faiz, A.H.M. Mustafizur Rahman, K.R. Islam (Eds.), Climate Change and Food Security in South Asia, XXII (2011), p. 600.

Ruddiman, W. F. (2003). The Anthropogenic greenhouse era began thousands of years ago. *Climatic Change, 61*(3), 261–293.

https://doi.org/10.1023/B:CLIM.0000004577.17928.fa

Tao, F., Yokozawa, M., Liu, J., & Zhang, Z. (2009). Climate change, land use change, and China's food security in the twenty-first century: an integrated perspective. *Climatic Change*, *93*(3–4), 433–445. https://doi.org/10.1007/s10584-008-9491-0

An Innovative Solution

American Museum of Natural History. (2020). *Vera Rubin on Dark Matter: A Factor of Ten.* American Museum of Natural History. https://www.amnh.org/learn-teach/curriculum-collections/cosmic-horizons-book/vera-rubin-dark-matter

Anderson, K., & Gaston, K. J. (2013). Lightweight unmanned aerial vehicles (UAVs) in ecological monitoring and biodiversity assessment. *Biodiversity and Conservation, 22*(4), 887–902. https://doi.org/10.1007/s10531-013-0451-7

Ashby, M. F. (2009). *Materials selection in mechanical design*. Butterworth-Heinemann.

Braun, D. R., Plummer, T., Ditchkoff, S. S., Ferraro, J. V., Maina, D., Bishop, L. C., & Potts, R. (2009). Raw material quality and Oldowan hominin toolstone preferences: Implications for early technological and cognitive evolution. *Journal of Archaeological Science, 36*(9), 1925–1933.

Bohm, D. (1980). Wholeness and the implicate order. Routledge.

Carroll, S. (2010). *From eternity to here: The quest for the ultimate theory of time*. Dutton.

Cohen, A. E., et al. (2012). Biomimicry for designing energy-efficient systems: The inspiration from photosynthesis. *Energy & Environmental Science, 5*(8), 8987-8997. https://doi.org/10.1039/c2ee03318k

Davies, P. (2006). *The Goldilocks enigma: Why is the universe just right for life?*. Houghton Mifflin.

Diamond, J. M. (1998). *Guns, Germs and Steel : a Short History of Everybody for the Last 13,000 Years*. Vintage.

Diem-Lane, A. (2016). Spooky physics: The illustrated edition (p. 17). MSAC Philosophy Group.

Dubrov, A. P., & Pushkin, V. N. (1983). *Parapsychology and contemporary natural science*. Moscow.

Dunbar, R. I. (1998). The social brain hypothesis. *Evolutionary Anthropology, 6*(5), 178–190.

Fitzpatrick, R. (2013). *Introduction to electrodynamics* (4th ed.). Pearson Education.

Griem, H. R. (2012). *Principles of plasma spectroscopy*. Cambridge University Press.

Guth, A. H. (1981). Inflationary universe: A possible solution to the horizon and flatness problems. *Physical Review D, 23*(2), 347–356. https://doi.org/10.1103/PhysRevD.23.347

IPCC. (2005). *Carbon dioxide capture and storage: Special report of the Intergovernmental Panel on Climate Change*. Cambridge University Press. https://doi.org/10.1017/CBO9780511546013

Kaplan, R., & Kaplan, S. (1989). *The experience of nature: A psychological perspective*. Cambridge University Press.

Keepin, W. (2024). *EP 1: The Inner Net of the Heart*. PWPA Los Angeles. https://www.pwpala.org/ep1-the-innernet-of-the-heart

Koch, C., & Bansal, S. (2014). Nature as a source of inspiration in the development of scientific understanding and technology. *Nature Reviews Physics, 1*(5), 1-11. https://doi.org/10.1038/s41567-019-0636-x

Kourtzi, Z., & Kanwisher, N. (2000). Cortical regions involved in perceiving object form. *Journal of Neuroscience, 20*(9), 3310–3318.

Kuntzelman, E. & Robinson, J. (n.d.). The Holomovement. A Whole World View for Social and Spiritual Transformation. *Kosmos Journal, 21*(3).

Lal, R. (2015). Restoring soil quality to mitigate soil degradation. *Sustainable Agriculture Reviews, 17*, 213–251.

Lal, R. (2020). Managing soils for resolving the conflict between agriculture and nature: The hard talk. *European Journal of Soil Science, 71*, 1–9. https://doi.org/10.1111/ejss.12857

Lal, R. (2023). Farming systems: Approaches and challenges for sustainability. *Farming Systems, 1*(100002). https://doi.org/10.1016/j.farsys.2023.100002

Lee, S., Ibarra, R., & Ihm, H. (2023). Yoga and Acupuncture for Depression and Prospective Side Effect of Psychiatric Medication. *1st International Conference of HyoJeong Academy 2023*, 40-42.

Linde, A. (1982). A new inflationary universe scenario: A possible solution of the horizon, flatness, homogeneity, isotropy, and primordial monopole problems. *Physics Letters B, 108*(6), 389–393. https://doi.org/10.1016/0370-2693(82)91219-9

Liu, C., & Liao, C. (2019). Artificial photosynthesis: A new frontier in the global carbon management. *Science, 363*(6426), 825–828. https://doi.org/10.1126/science.aav6508

Lund, J. W., & Boyd, T. L. (2016). Geothermal direct use: A unique resource for sustainability. *Geothermal Resources Council Transactions, 40*, 109–115.

Machín, A., Cotto, M., Ducongé, J., & Márquez, F. (2023). Artificial Photosynthesis: Current Advancements and Future Prospects. *Biomimetics*, *8*(3), 298. https://doi.org/10.3390/biomimetics8030298

Malkus, W. V. R., & Veronis, G. (1958). Convection and heat transfer in fluids. *Journal of Fluid Mechanics, 4*(3), 383–394. https://doi.org/10.1017/S0022112058000379

McLeod, E., Salm, R., Green, A., & Almany, J. (2008). Designing marine protected area networks to address the impacts of climate change. *Frontiers in Ecology and the Environment, 7*(7), 362–370. https://doi.org/10.1890/070211

McMichael, A. J., et al. (2006). Climate change and human health: Present and future risks. *The Lancet, 367*(9513), 859–869. https://doi.org/10.1016/S0140-6736(06)68079-3

Mitsch, W. J., & Gosselink, J. G. (2015). *Wetlands* (5th ed.). Wiley.

Moon, S. & Eu, H. (2006). *Exposition of the Divine Principle.* Holy Spirit Association for the Unification of World Christianity.

Mujumdar, A. S. (2013). *Handbook of industrial drying* (4th ed.). CRC Press.

NASA. (2023). *Dark Matter* - NASA Science. Science.nasa.gov. https://science.nasa.gov/mission/roman-space-telescope/dark-matter/

Nobel Prize. (2022). *Press Release*. The Nobel Prize. https://www.nobelprize.org/prizes/physics/2022/press-release/

Palinkas, C.M., Bolton, M. C., & Staver, L.W. (2023). Long-Term Performance and Impacts of Living Shorelines in Mesohaline Chesapeake Bay. *Ecological Engineering, 190*, p. 106944. https://doi.org/10.1016/j.ecoleng.2023.106944.

Pascalis, O., de Haan, M., & Nelson, C. A. (2002). Early experience with faces and perceptual narrowing. *Developmental Science, 5*(3), 379–387.

Peebles, P. J. E. (1993). *Principles of physical cosmology*. Princeton University Press.

Pezzotti, G., et al. (2010). Spider silk as inspiration for strong and flexible materials. *Nature Materials, 9*(9), 712-717. https://doi.org/10.1038/nmat2824

Rees, M. (1999). *Just six numbers: The deep forces that shape the universe*. Basic Books.

Rockström, J., et al. (2009). A safe operating space for humanity. *Nature, 461*(7263), 472-475. https://doi.org/10.1038/461472a

Salt, D. E., Smith, R. D., & Raskin, I. (1998). Phytoremediation. *Annual Review of Plant Biology, 49*(1), 643–668. https://doi.org/10.1146/annurev.arplant.49.1.643

Lee, S. (2014). *New Essentials of Unification Thought: Head-Wing Thought*. Unification Thought Institute.

Smolin, L. (1997). *The life of the cosmos*. Oxford University Press.

Taiz, L., Zeiger, E., & Møller, I. M. (2015). *Plant physiology and development* (6th ed.). Sinauer Associates.

Tompkins, P., & Bird, C. (1973). *The secret life of plants*. Harper & Row, Publishers.

US EPA. (2019, June 4). *How Do Wetlands Function and Why Are They Valuable? | US EPA*. US EPA. https://www.epa.gov/wetlands/how-do-wetlands-function-and-why-are-they-valuable

Vogel, S. (2003). *Comparative biomechanics: Life's physical world*. Princeton University Press.

Wilber, K. (Ed.). (1985). *The holographic paradigm and other paradoxes*. Shambhala/New Science Library.

Zeilinger, A. (2010). *Dance of the Photons*. Farrar, Straus and Giroux.

Zhang, Q., Hu, Y., Shao, M., Jia, X., & Wei, X. (2024). Revegetation re-carbonizes soil: Patterns, mechanisms, and challenges. *Fundamental Research*. https://doi.org/10.1016/j.fmre.2024.06.004

Epilogue

Jourdain, P. E. B. (1915). NEWTON'S HYPOTHESES OF ETHER AND OF GRAVITATION FROM 1679 TO 1693. *The Monist*, *25*(2), 234–254. http://www.jstor.org/stable/27900530

Kepler, J. (1619). Harmonices Mundi. As quoted in J. V. Field, Kepler's geometrical cosmology (p. 123). (1988).

Lee, S. (2014). *New Essentials of Unification Thought: Head-Wing Thought*. Unification Thought Institute.

Galilei, G. (1615). *Letter to the Grand Duchess Christina of Tuscany*. Stanford University. https://web.stanford.edu/~jsabol/certainty/readings/Galileo-LetterDuchessChristina.pdf

Made in the USA
Monee, IL
03 November 2026

67385322R00089